oranges & lemons

oranges
& lemons
coralie dorman
southwater

This edition is published by Southwater
Southwater is an imprint of Anness Publishing Ltd
Hermes House, 88–89 Blackfriars Road, London SE1 8HA
tel. 020 7401 2077; fax 020 7633 9499
www.southwaterbooks.com; info@anness.com

This edition distributed in the UK by The Manning Partnership Ltd
tel. 01225 478 444; fax 01225 478 440; sales@manning-partnership.co.uk

This edition distributed in the USA and Canada by National Book Network
tel. 301 459 3366; fax 301 429 5746; www.nbnbooks.com

This edition distributed in Australia by Pan Macmillan Australia
tel. 1300 135 113; fax 1300 135 103 customer.service@macmillan.com.au

Publisher: Joanna Lorenz
Managing Editor: Linda Fraser
Editor: Joy Wotton
Photographer: Nicki Dowey
Home Economist: Carol Tennant
Hand Model: Kirsty Fraser
Stylist: Helen Trent
Designer: Nigel Partridge
Additional Photography: Edward Allwright, Steve Baxter, Nicki Dowey, James Duncan, Gus Filgate, Ian Garlick, Michelle Garrett, Amanda Heywood, Janine Hosegood, David Jordan, Dave King, Don Last, William Lingwood, Thomas Odulate, Craig Robertson and Sam Stowell
Additional Recipes: Catherine Atkinson, Alex Barker, Angela Boggiano, Jacqueline Clarke, Carole Clements, Trish Davies, Roz Denny, Matthew Drennan, Joanna Farrow, Christine France, Yasuko Fukuoka, Brian Glover, Nicola Graimes, Christine Ingram, Lucy Knox, Sue Maggs, Sally Mansfield, Maggie Mayhew, Norma Miller, Sallie Morris, Jennie Shapter, Anne Sheasby, Hilaire Walden, Kate Whiteman, Elizabeth Wolf-Cohen and Jeni Wright

1 3 5 7 9 10 8 6 4 2

Previously published as part of a larger compendium, *The Citrus Cookbook*

NOTES

Bracketed terms are intended for American readers.
For all recipes, quantities are given in both metric and imperial measures and, where appropriate, measures are also given in standard cups and spoons. Follow one set, but not a mixture, because they are not interchangeable.
Standard spoon and cup measures are level. 1 tsp = 5ml, 1 tbsp = 15ml, 1 cup = 250ml/8fl oz
Australian standard tablespoons are 20ml. Australian readers should use 3 tsp in place of 1 tbsp for measuring small quantities of gelatine, flour, salt, etc.
Medium (US large) eggs are used unless otherwise stated.
The very young, the elderly, pregnant women and those in ill-health or with a compromised immune system are advised against consuming raw eggs or dishes and drinks containing raw eggs.

CONTENTS

FRUITS OF THE SUN

Ever since their earliest cultivation, oranges and lemons have been highly valued for their unique aroma and flavour. They have proved an important trading commodity, and have inspired a wealth of art and architecture, as well as providing a source of scientific study. Today, when almost 50 different cultivars of orange are grown around the world, it is hard to believe that there was a time when these fruits were unknown outside Asia.

First fruit

The bitter orange has been grown in southern China for thousands of years and this fruit probably originated there or in India. The first oranges to reach Europe would also have been bitter: there is evidence that the fruit was introduced to Spain by the Moors, and that it was in cultivation in Seville by the end of the 12th century, and in Italy by the 13th century.

The familiar Italian or Mediterranean lemon is actually a cross between a citron, the Indian lime, and another unknown member of the citrus family. It is believed to originate from the Punjab region of India and Pakistan. In the 4th century AD, lemons were brought to southern Europe by Arabs, and they were soon extensively cultivated in Spain, Sicily and North Africa.

Below: Every part of the Seville orange is highly valued, including the leaves, blossoms and fruit. Drawing from Kew Gardens Library, 19th century.

The name "orange" comes from the ancient Indian word *narayam*, meaning "perfume within". Arabs called the fruit *narandj* and in Italy the word became *arancia*. It is thought that, during medieval times in France, the fruit first shared the name of the town of Orange, where there was a large trade in the fruit. The name "lemon" is derived from Persian, via Arabic and French; the Chinese name for lemon is *li ming*. Unusually, there is no Latin word for lemon, although there is some evidence that the Romans used the lemon as a condiment, in medicine and for decoration. They may also have grown the fruit themselves.

Status symbols

Lemons were scarce in northern Europe during the Middle Ages and were an extreme luxury. Also a rarity, oranges became synonymous with wealth: the Italian Medici dynasty chose five oranges as its coat of arms. The fruits started to appear in art: Giuseppe Arcimboldo (1527–1593) produced extraordinary "portraits" of people composed entirely of different fruits; winter shows an old bearded man sprouting leaves and lemons. In France, lemon juice was used by the fashion-conscious to make the lips red and the complexion pale. Citrus fruits remained a symbol of prosperity until the 19th century. Gardeners had realized, ever since the invention of glass, that they could build structures in which to shelter plants from warmer climates, and the first orangery may have been that erected around 1600 by Henri IV of France, at the Tuileries; his successor Louis XIV built the most magnificent orangery in Europe at Versailles. The practice spread to Italy, and in the 19th century was taken up by gardeners and architects in Britain, where they created extravagant glasshouses which could be heated in the winter, providing somewhere for oranges to be grown and for the fashionable to stroll.

Above: Loading citrus fruit at San Antonio, Paraguay, 1891.

Bittersweet

Most oranges in Europe were bitter until the 19th century, although Vasco da Gama had probably brought back the first sweet oranges after his voyage to the Cape of Good Hope in 1498. Certainly the early orange varieties were far too bitter – and precious – to eat raw, so they were made into preserves, and prized for their perfumed flowers.

Although the Portuguese specialized in cultivating the bitter orange, patiently developing new varieties, most of the sweet oranges we eat today probably have their origin in trees Portuguese farmers imported from China in the middle of the 17th century. These early varieties were sold by the likes of "pretty, witty Nell Gwynne" as a popular and exotic snack in London's Drury Lane Theatre. At that time, marmalade was a stiff, candied paste that could be cut into squares as a sweetmeat. The world's most popular eating orange, Valencia, was sent in the 1860s from the Azores Islands, near Portugal, to Thomas Rivers of Hertfordshire, England, who cultivated it for orangeries in grand houses. Originally called Rivers Late, it was renamed Valencia Late in

1887, because an expert from Spain thought that it was similar to an orange that was grown in Valencia.

For culinary purposes, though, the bitter Seville (Temple) orange is still highly prized. In the 17th century, its season extended from early November to April. In Europe it now lasts a mere four weeks, but the Seville orange is still considered by experts – and seasoned preserve makers – to be the only variety to use for marmalade. Every part is highly valued, including the flowers, leaves and rind. The blossoms are turned into intensely perfumed orange water or orange blossom water, particularly in the Middle East.

The New World

In 1493 Christopher Columbus took lemon seeds to the New World, where the fruit was unknown. The Portuguese introduced lemons to Brazil, and the Brazilians, in turn, took them to Australia, planting the first trees in 1788. Scurvy was rife on these early colonist ships, but it wasn't until the late 18th century that it was realized that the disease, caused by a lack of vitamin C, could be cured by eating citrus fruits. Columbus is credited with introducing the orange to America – oranges were first cultivated in Florida in the 16th century, and they were introduced to California 200 years later.

The orange-growing industry became solidly established towards the end of the 19th century and the United States now grows about one-sixth of the world's supply. The other main producers of oranges are China, Spain, Greece, Italy, Morocco, Turkey, Israel, Australia and South Africa.

Origins of liqueurs

Both oranges and lemons are used in the production of alcoholic drinks. Dutch settlers in the Caribbean island of Curaçao started the tradition, when they flavoured white rum with the bitter peel of green oranges, and the technique was eventually adopted by the French. In 1849, the brothers Edouard and Adolphe Cointreau launched a product called Triple Sec White Curaçao, which eventually became known as Cointreau.

Above: A view of the orangery in Lord Burlington's garden, Chiswick, England, 18th century.

France's other famous orange liqueur, Grand Marnier, was devised in 1880 by Louis-Alexandre Marnier. Elsewhere, lemons were used in Limoncello in Italy and Limonnaya vodka in Russia.

A taste of the sun

All citrus fruits contain vitamin C, and oranges contain vitamin A and many minerals. Lemons are a good source of bioflavonoids, cleansing the blood, and the juice is a well-known antiseptic.

Today, virtually every cuisine is touched by the refreshing zesty tang of citrus fruit. Oranges are the world's favourite fruit, while lemons provide undertones in hundreds of dishes.

Every country has its own traditional ways of using oranges and lemons, from the Moroccan practice of preserving lemons in salt, to Greek egg-and-lemon soup, and British orange marmalade – and the recipes in this book show just how versatile these fruits can be.

ORANGES

Arguably the world's favourite citrus fruit, oranges are not always orange. They can be green, yellow or dappled with red. There are hundreds of varieties, ranging from fruits the size of a tennis ball to those that are almost as big as a football. Some have honey-sweet flesh, while others are so bitter as to be quite inedible.

All citrus peel contains essential oils, and orange peel is used in perfume-making and cosmetics, as well as cooking. Depending on the variety, oranges grow well in tropical, semi-tropical and sub-tropical regions. They ripen on the tree, and do not continue to ripen after picking, so have excellent keeping qualities.

Varieties

Oranges fall into two principal groups: sweet, which can be eaten raw, and bitter, which cannot.

Sweet oranges, *Citrus sinensis*

Blond or common oranges These are pale-skinned winter varieties such as Jaffa or Shamouti (the easiest to peel). A variety called Pineapple is grown in Florida, mainly for its juice as it can be rather pippy, and Natal comes from Brazil. Valencia is the world's favourite eating orange and is grown in all the orange-producing countries. The fruits vary in size from medium to large, with reasonably thin, finely marked rind. Valencias are easy to peel, and they have delicious juicy flesh.

Blood or pigmented oranges If it's colour you're after, these are the oranges to use, especially for vibrant sorbets (sherbets) and similar iced desserts. These fruits have names such as Sanguina, Sanguigna and Sanguine, all names which derive from *sanguis*, the Latin word for blood. They have red-tinged skin and golden or ruby red flesh. The red pigment develops naturally when the citrus trees grow where the temperature at night is low; cold storage can effect a similar change.

Bittersweet blood oranges originated in Malta or Sicily, and the fruit is used in sauce Maltaise, a mayonnaise named after the native juicy but sour variety of blood orange. Moro is a variety grown mainly in Italy, particularly in Sicily.

Navel oranges These take their name from the small orange that grows inside the apex of the main fruit, a bit like a belly button. The navel, which is the earliest orange to mature, is seedless, sweet and juicy and great for eating in the hand. The most popular varieties are the Washington or Bahia navel, Cara Cara, and Lane Late from Australia, or its renamed Spanish version, Navelate. The thick skin makes good candied peel, used in fruit cakes and mince pies.

Left: Bitter Seville oranges are only available for a short season, but they do freeze well.

Left: The very best orange marmalade, many people consider, is made using Seville oranges.

Sugar or acidless oranges These varieties are not cultivated on a vast scale and have a very low acid content and a bland, rather insipid flavour. They are seldom exported, but are usually grown for local consumption. Sugar oranges grow mainly in the Mediterranean and in South America, where they are known by a range of names, which may refer to their sweetness, such as Lima (Brazil), Dolce (Italy) and Sucreña (Spain).

Bitter oranges, *Citrus aurantium*

The first variety of orange, Seville oranges – known as Bigarade in France and Temple in the USA – taste bitter, owing to a compound called neohesperidin. You can't eat them raw, but they are ideal for use in marmalade. In industry, they feature in liqueurs and some soft drinks, and the flowers from special varieties called bouquetiers are used to make oil of neroli, used in perfumery. In Europe, Sevilles make a fleeting appearance in the markets in January; fortunately they freeze well, as does their juice, so if you have enough freezer space you can store them and make marmalade when you feel like it. The rind is good dried.

Nutrition

Packed with vitamin C, which is highest at the start of the season in winter, oranges are also a source of vitamin A and folic acid, as well as calcium, potassium and phosphorus. The white pith contains pectin, which, as well as acting as a setting agent for marmalade, helps to lower cholesterol levels, and also bioflavonoids, which keep gums and blood healthy.

Below: Seedless navel oranges have very sweet juicy flesh and are excellent for juicing and eating.

Buying and storing

Oranges crop at different times, depending on variety and climate. You can buy them all year round, but winter is the optimum time. A heavy fruit will probably be full of juice, and the skin should be taut and blemish-free. Although they are an irritant when eating or preparing the fruit, pips (seeds) are not a sign of inferior quality; their presence just depends on the variety. The juice and pared or grated rind can be frozen.

Serving ideas

The simplest way to enjoy a fresh orange is to quarter it and then extract the flesh with your teeth. A less messy method is to separate the segments, remove as much white pith as possible and serve them, perhaps moistened with freshly squeezed orange juice or a dash of Cointreau or Curaçao. Oranges can also be chilled and sprinkled with a little sugar to bring out their juicy sweetness.

The fruit makes an excellent flavouring for cakes, particularly rich fruit cakes, plain Madeira cakes and crisp biscuits (cookies). The decorative segments and slices look equally good in home-made fruit jellies, or as a caramelized topping for an orange flan. Oranges make very good sorbets (sherbets), ice creams and savoury or sweet sauces, especially when teamed with brandy or an orange liqueur as in the classic dessert crêpes Suzette.

Oranges are equally at home in a savoury setting. Orange rind is used to flavour stews in both French and Chinese cuisine, while segments are used in canard à l'orange. Juice and rind are excellent partners in marinades, stuffings or sauces for meat, poultry and fish, and they make a refreshing salad dressing; orange segments are classically combined with beetroot (beet) and chicory (Belgian endive). Use the juice and rind to perk up mild, creamy cheeses and to complement sweet red onions and glazed carrots.

Commercial uses

Oranges are highly versatile so it will come as little surprise that they are used in many products, from freshly squeezed orange juice and cordials through ready-made fruit salads and cakes, to marmalades, curds and dressings.

Marmalades

These days, commercial marmalades may contain every type of citrus fruit, with ginger, vanilla and other spices joining spirits, such as brandy and whisky, to create exciting new flavours. Given the choice, however, many people still plump for a traditional recipe, and thick-cut Seville orange is one of the best.

Liqueurs and cordials

Curaçao Dutch settlers on the Caribbean island of Curaçao first made this white rum-based liqueur, also known as Triple Sec. It is flavoured with bitter green oranges (not a bitter variety, just unripened), and brandy is often used instead of rum. Curaçao is always orange-flavoured and comes in a variety of colours – red, yellow, blue and vivid green – the clear type is always Triple Sec.

Cointreau This is a Curaçao with a brandy base. It is very sweet and made with a blend of the Caribbean bitter green oranges and sweeter varieties from France.

Grand Marnier This branded Curaçao blends unripened oranges with the finest Cognac, and is aged in barrels. Grand Marnier has a highly refined, mellow flavour and the sweetness is tempered by the Cognac.

Aurum The name of this brandy-based Italian liqueur hints at gold, which it may once have contained. Today, however, the glorious golden colour is the result of a mixture of infused orange peel and whole oranges, enhanced by the addition of fragrant saffron.

Left: Grand Marnier is made from bitter green oranges and fine Cognac. Louis-Alexandre Marnier first hit upon the idea for the drink after tasting bitter oranges in Haiti in 1880.

LEMONS

Lemons are the secret ingredient in many dishes. A squeeze of lemon transforms a plain poached fish into a meal to remember, cuts through the sweetness of a fruit salad. Without it, avocados and artichokes would turn brown. With their fresh, tart flavour, lemons are an excellent single flavouring and enhance countless other ingredients.

Lemons thrive in a range of climates, but most are grown in sub-tropical regions, as the humidity in hotter countries encourages pests and diseases. Unlike other citrus fruits, lemon trees blossom several times during the year and trees contain fruit of various ages. Lemons do not ripen once picked and last for up to two weeks.

Varieties

Distinguishing varieties of lemon can be extremely tricky. There is less difference than with oranges, and the shape and thickness of the skin vary according to the time of year. The skin and pith can be thick or thin, the skin can be smooth or bumpy, and the juice may be mouth-puckeringly sour or almost sweet.

Eureka Originally cultivated in Europe, and still commonly grown in Spain, this variety is today a favourite in Australia, South Africa, Argentina and Israel. Fairly thin-skinned and very juicy, it is also popular in California, where it was introduced from Sicily in 1858.

Right: Varieties, like these Jambhiri lemons, are rarely named in stores.

Femminello Commune This variety comprises around three-quarters of the Italian lemon crop, but is also grown in Argentina and Turkey. In Italy, it crops four times a year and each season's crop has a different name: Primofiori (September–November), Limoni (December–May), Bianchetti (April–June), and lastly Verdilli (June–August). Femminello Commune lemons have medium skin and pith, but a lower than average juice content.

Fino or Primofiori This smooth-skinned Spanish variety is relatively small, and is rounded or oval. In the northern hemisphere, it crops between October and February.

Lisbon Originally a Portuguese variety, this was developed in Australia, where it is still grown successfully. Lisbon lemons are also grown in California and South America. The juicy fruits are quite large, with fairly thin, slightly rough rind. They crop during the winter and into early spring.

Verna This fruit accounts for about 60 per cent of the Spanish summer crop and has a rough, rather thick rind.

Nutrition

As well as vitamin C, lemons are an excellent source of bioflavonoids, which have antioxidant properties that help maintain the immune system and protect against disease, including some forms of cancer. They also contain small amounts of potassium, calcium and phosphorus. Lemons are also said to ease rheumatism.

Left: Smooth-skinned lemons are the best when it comes to juicing.

Buying and storing

Growers talk of nipples (the stalk end) and necks (where the lemon elongates) but, apart from minor differences in these areas, the only feature that distinguishes lemons at the market is the roughness or smoothness of the skin. Lemons are seldom labelled as to variety; the most you can expect is a clue to the country of origin. Until you cut it open, you cannot be sure whether your choice is juicy, or has thin skin and pith, or thick skin with up to 1cm/½in of pith. Look for sprightly, taut skin, with a clear, bright yellow colour. Lemons gradually become paler and less juicy with ripening.

Where possible, always buy unwaxed lemons, sometimes described as untreated, or organic lemons. The wax makes the fruit shine and contains an antifungal preservative, biphenyl. It is considered harmless, but if you intend to grate or pare the rind, thoroughly scrub a waxed lemon first. This will also remove any pesticide residue.

Serving ideas

Many dishes, both sweet and savoury, benefit from the distinctive tang of lemon, from green salads with lemon vinaigrette to lemon meringue pie. The

Above: Lemon products include fizzy soft drinks, pre-squeezed lemon juice and herbal tea bags.

fruit is seldom served solo, except when preserved or transformed into fresh lemonade, but it is used primarily to enhance or flavour other foods.

Lemon juice has numerous uses; it tenderizes meat, gives fresh cream a soured taste, and prevents fruit and vegetables – especially apples, celeriac, artichokes, avocados and potatoes – from oxidizing and browning. It also makes a great marinade for both fish and shellfish, but do not marinate these foods for over an hour or the juice will start to "cook" or denature the protein. The juice can feature as a principal flavouring, as in lemon cheesecake, mousse and sorbet (sherbet), or as a complementary flavour, such as in lemon mayonnaise or cake frosting.

It is lemon skin that contains the really powerful flavour, however. If you rub a lump of sugar over a lemon it absorbs the aromatic oil. Use the sugar in a dessert sauce or cocktail, to add a subtle citrus flavour. Grated or pared lemon rind tastes wonderful in cakes, desserts and even savoury dishes, such as beef, lamb, pork or chicken casseroles. When paring or grating lemons, avoid including any of the white pith, which tastes unpleasantly bitter. The pith is useful, however, in helping jam and preserves to set.

Preserved lemons feature in both sweet and savoury recipes. Sliced or quartered lemons, preserved in salt and stored in olive oil, are widely used in North Africa, in chicken and fish dishes. Candied slices or half slices may be used to decorate desserts. Lemon slices or wedges are classic garnishes for hot and cold fish and shellfish dishes. They are also natural partners for chicken and veal and the favourite decoration for cocktails.

Finely grated lemon rind may be sprinkled over both sweet and savoury dishes and, combined with flat leaf parsley and garlic, it forms the Italian garnish, gremolata.

Commercial uses

The main commercial uses of lemon have been in drinks, spirits and liqueurs: choose from lemon tea, lemon cordial or traditional, old-fashioned lemonade, and use lemon liqueurs to enliven cocktails or serve with coffee. Pre-squeezed lemon juice is also available for culinary use.

Lemon drinks

Limoncello This Italian liqueur is made from a base alcohol steeped with lemon rind and sugar. Add to a fruit salad or drizzle over raspberry ice cream.

Liquore al limone or Cedro This sticky, sweet liqueur is made from the peel of lemons that grow around Italy's Amalfi coast. Served ice-cold, straight from the refrigerator or freezer, it makes a refreshing aperitif or digestif.

Limonnaya Fruit flavourings are quite common for vodka. Limonnaya is a leading lemon-flavoured brand.

Lemon barley water A traditional English cordial, this is made from ground barley, water, lemon rind and juice, sweetened with sugar or honey. Its popularity has declined, but it remains a good choice at picnics or after sport.

Lemon cordial In Britain, the name is given to a wide range of concentrated lemon drinks that are mixed with water. Some are full of sugar or artificial sweeteners, but better quality cordials are useful for making long, cold drinks for summer afternoons. American cordials are alcoholic.

Right: Limonnaya, a popular brand of lemon-flavoured Russian vodka

Far right: Liquore al limone

Equipment

Although you can squeeze an orange with your bare hands, it is much easier with equipment designed especially for the job, whether this be a simple reamer or a sophisticated juicer. There are several other inexpensive gadgets that will prove handy in preparing citrus fruit.

Vegetable peeler

Use a vegetable peeler to pare away thin strips of citrus rind, leaving the white pith behind. There are two types to choose from – the fixed blade peeler, which demands little dexterity in the hands of the user, but doesn't cut quite such a fine strip, and the swivel-blade peeler, which is sharper and more flexible and therefore more precise.

Left: Fixed- and swivel-blade vegetable peelers

Cannelle knife (zester)

This handy tool has a tooth-like blade, which will shave citrus rind into ribbons or into fine julienne strips, leaving all the bitter white pith behind.

Citrus zester

Scrape the five-hole blade of the citrus zester over oranges or lemons and the peel will come off in fine curly ribbons. A useful combination tool, which is both a citrus zester and a cannelle knife, is also widely available.

Above: Cannelle knife (left) and combination citrus zester and cannelle knife (right)

Citrus reamer

Below: Citrus reamer

This curvy wooden device is simple but effective, letting you squeeze the juice straight into a bowl or pan. You simply insert the pointed end into a halved lemon or orange, and twist. Its only drawback is that you can't catch the pips (seeds), but if you hold the halved fruit in one hand with slightly opened fingers, and carefully twist the reamer with the other, you should be able to catch them. Alternatively, squeeze the juice into a jug (pitcher), then strain it into the bowl or pan to catch the pips.

Below: Lemon squeezer

Lemon squeezer

Most of us have one of these tucked away in the kitchen. The basic design has barely changed in generations, but lemon squeezers now come in a variety of shapes, sizes and materials – glass, plastic or chrome. The business end of the utensil is a dome on to which you push a halved fruit, then twist to extract the juice. The pips are trapped in the rim and the juice either flows into a simple channel surrounding the dome, or into a bowl beneath. Some models of squeezer have interchangeable heads, which means that they can be used for different sizes of fruit.

Citrus press

If you want to squeeze the maximum amount of juice from the minimum quantity of fruit, this is well worth the investment. The simplest version is a hand-operated gadget with a geared lever mechanism. The halved fruit is placed inside, and as you pull the lever forward, the juice is squeezed into the integral container.

Below: A hand-operated citrus press

Lemon tap

A simple gadget that turns citrus fruit into a "juice jug". If only a small amount of juice is required, leave the tap inserted in the fruit to keep the juice fresh between squeezings.

Electric juice extractor

These machines extract juice from oranges, lemons and other fruits. They take up a lot of space but are invaluable for lovers of fresh fruit juice.

Box grater

Choose a stainless steel grater with as many different grating surfaces as possible, preferably with a handle for a better grip. Make sure it stands firmly on the work surface or in a bowl. Some models include a flat blade for thinly slicing lemons and oranges, too.

Microplane grater

A real revolution when it comes to grating. The blade is actually a chemically sharpened grating surface that will take the outer rind off a lemon in next to no time, and with very little effort. It comes with an extremely comfortable soft grip handle, and it is very easy to brush off every last bit of grated rind from the back when it comes to cleaning.

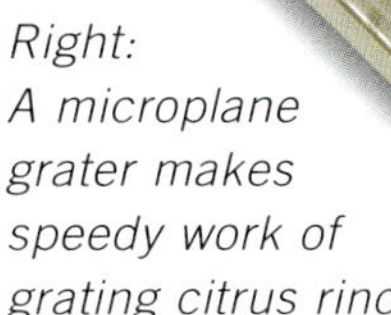

Right: A microplane grater makes speedy work of grating citrus rind.

Left: Box grater

Funnel

Essential for pouring hot marmalades and jellies into jars. Stainless steel funnels withstand heat better than plastic ones.

Preserving pan

Especially designed for preserving, this is a double-handled pan with a thick, heavy base to prevent burning. The best preserving pan will also be wide and fairly shallow, which makes for rapid, highly effective boiling. Choose a pan that is made from stainless steel or pristine (that is, not cracked) enamel: avoid aluminium, which will taint the fruit. Preserving pans also come with a non-stick lining.

Right: Sugar thermometer

Below: Preserving pan with a square of muslin (cheesecloth) for straining marmalades

Tips for grating citrus rind

- Grate citrus rind on to a sheet of foil or baking parchment. You can then tip it into a bowl or pan.
- Use a small pastry brush or even a new toothbrush to remove the grated rind from the grater.
- Make sure that you only grate the coloured rind and not the bitter-tasting white pith.

Sugar thermometer

The most accurate way to check the temperature of jams, jellies and syrups is to use a thermometer. It will help you determine when the setting point has been reached. A thermometer with a plastic handle will prove easy to hold in hot liquids.

Basic Techniques

Preparing lemons and oranges for use in cooking, to eat fresh or for decoration, is not hard but it does help to master a few techniques. The cheering colours of these fruits make them ideal for use in decoration, and even simple touches enliven dishes. The following techniques are shown with the fruit most likely to demand that particular skill, but they can be used with either fruit.

Peeling and segmenting fruit

1 Using a serrated knife, cut a small slice off both the top and bottom of the orange or lemon to expose the flesh and make it easier to remove the peel.

2 Holding the fruit in one hand, cut off the peel in one continuous strip, starting at the top and cutting around the fruit in a circular motion, using a gentle sawing action and making sure that you remove every scrap of white pith. Alternatively, stand the orange or lemon on a plate and cut off the peel in strips, working from top to bottom.

3 Holding the fruit over a bowl to catch all the juices, remove each segment in turn by cutting down either side of the membrane and easing it away.

4 When you have removed all the segments, squeeze the membrane over the bowl to extract the remaining juice.

Zesting a lemon or orange

If you intend to use the rind or peel of a lemon or orange in cooking, choose an unwaxed or organic fruit or scrub waxed fruit thoroughly first.

1 To zest a lemon or orange, hold the fruit very firmly in one hand. Scrape a lemon zester down the length of the fruit.

2 If you need the zest to be even finer, brush it into a neat pile on a chopping board, hold a large chef's knife at right angles to the board and rock it back and forth to chop the zest finely.

Grating rind

Freshly grated orange or lemon rind will enhance many dishes.

Hold the fruit against the fine side of the grater and work it up and down, taking off only the zest and not the pith.

Paring and cutting julienne strips

Julienne is the term used for fine and delicate strips or sticks.

1 Use a vegetable peeler to remove long, wide strips of rind. Be careful not to include the bitter white pith.

2 Square off the ends of the strips, then cut them lengthways. Use immediately or set aside, covered with cold water.

Slicing oranges and lemons

Thin slices of unwaxed fruit will enhance homemade lemonade and other drinks.

Slice the fruit horizontally, using a sharp, serrated knife. Remove the pips (seeds), then cut each slice in half.

Making citrus wedges

A squeeze of lemon or orange adds subtle sharpness to many fish recipes.

Cut the fruit lengthways, in half, then into quarters and finally into thin wedges. The juice is directed down when squeezed.

For decorative wedges, cut the fruit in half crossways, cut in half through the stalk end and then cut into wedges.

Drying orange and mandarin rind

When dried, the peel of thin-skinned oranges and mandarins gives a wonderful flavour to meat stews and casseroles, and it can also be used in soups, vegetable dishes, sauces, salad dressings, desserts and cakes.

1 Wash unwaxed fruit thoroughly, then peel. Cut the peel into large pieces and place each piece on a board, outer side down. Use a small knife to scrape away every trace of white pith.

2 Cut the peel into thin strips, about 2.5cm/1in long, and spread these out on a tray. Cover loosely and leave to dry in an airy room for about one week, until they are completely dry and have darkened in colour. Alternatively, dry in an oven preheated to 110°C/225°F/ Gas ¼ for 10–12 hours, then leave until completely cold.

3 Tip the peel into a clean, screwtop jar, replace the lid and keep in a cool, dry place. Use the rind whole, crumble between your fingers or powder in a coffee grinder. One or two strips, or 5ml/1 tsp ground peel, will be enough to flavour most dishes.

Making a citrus bouquet garni

In Provence in the South of France, it is traditional to add a strip of citrus rind to the herbs in a bouquet garni, especially when it is to be used to flavour a beef stew or casserole. It would also go well in rich game and duck dishes.

1 Pare a broad strip of rind from an orange or mandarin. A clementine or satsuma could be used instead, but avoid very thin-skinned fruit. Remove any white pith from the rind.

2 Tie the rind in a neat bundle with a sprig each of fresh thyme and rosemary and a few stalks of fresh parsley. Add a bay leaf if you like. Leave the end of the string long enough to loop around the handle of a pan or casserole. This prevents the herbs from dispersing through the stew or casserole, making for easy retrieval before serving.

VARIATION

If a bouquet garni is to be used in fish or shellfish recipes, use lemon rind instead of orange or mandarin, and add a sprig of fresh tarragon and dill to the herb bundle. If you like, tie the bouquet with a thin strip of celery.

Making Citrus Decorations

Making caramelized citrus rind

The bold colours of lemons and oranges combine to make a pretty decoration on cakes and desserts, such as lemon cheesecake and orange flan. The caramelized rind is sweet, with an underlying citrus tang.

1 Using a swivel-bladed vegetable peeler, remove the rind from washed, unwaxed fruit in long, even strips. Using a small, sharp knife, slice into fine julienne strips.

2 Place the strips in a small pan of water, bring to the boil, then lower the heat and simmer for about 5 minutes. Drain, refresh under cold water and then drain again. Pat dry with kitchen paper. Line a grill (broiler) pan with foil and preheat the grill.

3 Spread out the rind flat on a sheet of baking parchment and dredge with icing (confectioners') sugar. Toss to coat, then tip on to the foil-lined grill pan. Grill (broil) for 2–3 minutes, or until the sugar just begins to melt, then leave to cool and harden. Pile on top of a cooled cake or dessert, dusting with more icing sugar, if you like.

Making citrus corkscrews

Hang these over the edge of ice cream dishes or use to decorate a cocktail flavoured with citrus juice.

1 Using a cannelle knife (zester), pare long strips of rind from a lemon or orange. The strips should be as long as it is possible to make them.

2 Pick up a cocktail stick (toothpick) and wind and twist the strip of citrus rind tightly around it. Slide the cocktail stick out, and you will be left with a little citrus corkscrew. Make more in the same way.

Making citrus loops

1 Cut a lemon in half lengthways, then cut each half into four equal segments. Holding one segment in your hand, use a sharp knife to ease away the flesh from the skin in a single piece, leaving about 1cm/½in still attached at the top.

2 Carefully cut away any bitter, white pith still remaining on the now bare piece of lemon peel, then fold it under the flesh segment to form a loop. Make three more lemon loops from the remaining segments.

Making decorative slices

A pretty slice of orange or lemon is an irresistible addition to many long, cool drinks or cocktails.

1 Using a cannelle knife (zester), pare strips of the rind at regular intervals.

2 Using a sharp, stainless-steel knife, cut the orange into thin, even slices.

Making citrus leaves

Two of these stylish leaves, placed side by side, make a dramatic garnish for a smoked salmon terrine.

Cut a V-shaped wedge from the side of a lemon. Cut the lemon lengthways in 5mm/¼in slices, following the curve created by removing the wedge.

Making citrus shells

Oranges and lemons look lovely filled with ice cream, sorbet (sherbet) or mousse. The shells of larger oranges are wonderful for individual salads, such as orange and beetroot (beet) or chicory (Belgian endive).

1 Cut the top off the fruit and set it aside to use as a lid. Loosen the flesh by running the blade of a small, sharp knife between the flesh and the skin. Scoop out the flesh with a teaspoon, keeping the shell intact, and set it aside for another dessert. Rinse the shell and leave it upside down to drain. Make more shells in the same way.

2 Pipe or spoon the filling into the fruit shells and replace the lids, setting them at an angle if you like. If the filling is ice cream or sorbet, wrap the filled fruit in clear film (plastic wrap) and freeze. If the filling is a mousse, cover loosely, chill and serve within the hour.

COOK'S TIP

To make shells more decorative, use a cannelle knife (zester) to make lengthways grooves in the skin before cutting off the top and hollowing out the fruit.

Making a citrus ice bowl

An ice bowl makes a glorious centrepiece when filled with little balls of citrus sorbet (sherbet) or ice cream. Pale colours are more effective than bold ones.

1 You will need two freezerproof glass or plastic bowls that will fit one inside the other, leaving a gap between them of about 2.5cm/1in all around. Add some crushed ice to the larger bowl to hold up the inner bowl, then tape them together with parcel tape.

2 Carefully fill the gap between the bowls with water. It should come almost to the top, and the smaller bowl should float slightly.

3 Slide thin slices of orange and lemon into the water between the two bowls. Use a skewer to push the slices of fruit into position. Carefully lift the pair of bowls into the freezer and freeze overnight, until required.

4 To unmould the ice bowl, peel off the tape. Put the bowls in another larger bowl half-filled with hot water, then pour a little hot water into the smaller bowl. Leave it there for about 30 seconds or so, then lift both bowls out of the water. Pour the water out of the smaller bowl and loosen the inner bowl with a thin, round-bladed knife.

5 Lift out the inner bowl, turn out the ice bowl and put it on a large plate. You can sprinkle a few extra slices of citrus fruit or curled rind around the edge if you like, or add a few flowers, such as nasturtiums. Fill with scoops of ice cream or sorbet and serve.

Old-fashioned lemonade

This home-made drink is both cooling and free from additives.

SERVES TWO

INGREDIENTS

- 2 lemons
- 75g/3oz/6 tbsp caster (superfine) sugar
- 2 egg whites
- 300ml/½ pint/1¼ cups boiling water
- 75ml/5 tbsp sherry (optional)
- ice cubes

1 Pare a few strips of rind from one of the lemons, and place in a pan. Cut both lemons in half and squeeze the juice.

2 Strain the juice to remove any pips (seeds), then add to the pan.

3 Add the sugar and egg whites and whisk until frothy. Whisk in the water for 2 minutes more, until the sugar has dissolved. Pour in the sherry, if using, a little at a time, whisking after each addition.

4 To serve, put several ice cubes into tall glasses and add the lemonade.

MARMALADES

Breakfast just wouldn't be the same without marmalade. The name comes from the Portuguese *marmelada*, which was a sweet, stiff paste made from quinces. In Tudor times, peach, apple, pear, lemon and orange pastes were all called marmalades. Resembling fruit cheeses, they were usually cut into pieces and presented in fancy boxes as a sweetmeat. By the 18th century, a clear, looser version existed, which was a precursor to today's preserves. The bitter Seville (Temple) oranges were the most popular fruits to use, and they remain a favourite today.

Seville orange marmalade

This classic recipe is made with the best oranges for the job: Sevilles.

MAKES ABOUT 4KG/9½LB

INGREDIENTS

- 1kg/2¼lb Seville oranges
- 1 unwaxed lemon
- 2.2 litres/4 pints/9 cups water
- 2kg/4½lb/9 cups preserving sugar

1 Wash the oranges and lemon and cut them into quarters. Remove the flesh, pips (seeds) and pulp and tie them in a square of muslin (cheesecloth). Shred the peel finely or coarsely, depending on how you like your marmalade.

2 Place the peel in a preserving pan and add the water and the muslin bag. Bring to the boil, then lower the heat and simmer for 1½–2 hours. About 15 minutes before the peel is ready, preheat the oven to 110°C/225°F/ Gas ¼. Put the sugar in a heatproof bowl and place it in the oven to warm.

3 Remove the muslin bag and hold it over the pan between two plates. Press to extract as much liquid as possible. Stir in the sugar until it has dissolved, then boil rapidly until setting point is reached. Pour into warm, sterilized jars and seal. Label when cold.

Orange and coriander marmalade

This marmalade has a sweet and spicy flavour that goes well with toast, but also with cold meats or hot sausages.

MAKES ABOUT 2.75KG/6LB

INGREDIENTS

- 1.6kg/3½lb oranges, washed
- 6 unwaxed lemons, washed
- 30ml/2 tbsp coriander seeds, roasted and coarsely crushed
- 3 litres/5 pints/12 cups water
- 1.6kg/3½lb preserving sugar

1 Cut the fruit in half, then squeeze and pour the juice into a preserving pan.

2 Scrape the pith from the citrus shells and tie it up, with the pips (seeds) and half the coriander, in a square of muslin (cheesecloth). Tie the bag to the handle of the pan so it dangles in the juice.

3 Slice the peel into shreds and add them to the pan with the water. Bring to the boil, then simmer for 1½ hours. Preheat the oven to 110°C/225°F/ Gas ¼. Put the sugar in a heatproof bowl and place it in the oven to warm.

4 Remove the muslin bag. Hold it over the pan and squeeze it between two plates to extract all the liquid. Stir in the sugar and the remaining coriander, then boil rapidly until setting point is reached. Skim off any scum, then leave for about 30 minutes. Pour into sterilized jars and cover with waxed discs. Leave to cool before sealing and labelling the jars.

Tips for a successful marmalade

The best way to test for setting point is with a sugar thermometer. When it registers about 105°C/ 220°F, the marmalade is ready.

To sterilize jars, preheat the oven to the highest setting. Wash the jars in hot, soapy water, then rinse. Stand them in the sink and pour over boiling water. Place them on a baking tray in the oven for 10 minutes.

Lemon and ginger marmalade

This classic combination of lemon and ginger produces a really zesty preserve, perfect when served on toast or scones, or you could try mixing it with a little soy sauce and using as a glaze for roast meat. It also makes an excellent cold remedy when stirred into boiling water with a dash of whisky!

MAKES 1.8KG/4LB

INGREDIENTS

- 1.2kg/2½lb lemons
- 150g/5oz fresh root ginger, peeled and finely grated
- 1.2 litres/2 pints/5 cups water
- 900g/2lb/4½ cups warmed sugar

1 Wash the lemons then cut into quarters and slice, reserving any pips (seeds). Tie the pips in a muslin (cheesecloth) bag with any fibrous ginger pieces left after grating. Place the lemons in a large pan with the grated ginger and water. Bring to the boil, then cover and simmer for 2 hours or until the fruit is very tender.

2 Remove the muslin bag and squeeze over the pan to release all the juice and pectin, then discard. Stir in the sugar over a low heat until completely dissolved, then increase the heat and boil for a further 5–10 minutes or until the marmalade reaches setting point.

3 Remove the pan from the heat and skim off any scum from the surface using a slotted spoon. Leave to cool for 5 minutes, and then pour into warmed sterilized jars. Seal and label, then store the jars in a cool, dark place.

Three fruit marmalade

You could use sweet oranges, such as Valencia or Navelina, in place of the traditional bitter Seville oranges, but the flavour of the marmalade would not be so intense and tangy. Most marmalades require all the pith to be included on the peel, but for this marmalade, don't have it too thick.

MAKES ABOUT 2.25KG/5LB

INGREDIENTS

- 2 unwaxed or well-scrubbed Seville oranges
- 2 unwaxed lemons
- 1 grapefruit
- 1.75 litres/3 pints/7½ cups water
- 1.5kg/3lb/6¾ cups preserving sugar

1 Wash the fruits and cut them in half. Squeeze all the fruit and pour the juice into a preserving pan. Tie the pips (seeds) and pulp securely in a square of muslin (cheesecloth). Tie the bag to the handle of the pan so that it dangles in the citrus juice.

Marmalade or jam?
The main difference between marmalade and jam is that the former invariably contains citrus fruit. Fruits such as oranges and lemons are used to make fruit curds or cheeses, but we seldom find them in jam, except in order to boost pectin levels. One exception, however, is a light orange jam, from Portugal, which makes a delicious spread.

2 Cut the citrus skins into thin wedges; scrape off and discard the membranes and pith. Cut the peel in slivers and add them to the pan with the water. Bring to simmering point and cook gently for 2 hours, or until all the peel is tender and the water has reduced by half.

3 About 15 minutes before the peel is ready, preheat the oven to 110°C/225°F/Gas ¼. Put the sugar in a heatproof bowl and place in the oven to warm.

4 Remove the muslin bag. Holding it over the pan, squeeze it between two plates to extract all the liquid. Stir in the sugar until dissolved.

5 Bring to the boil and boil rapidly until setting point is reached. Pour a small amount on to a chilled saucer. Chill for 2 minutes, then push the marmalade with your finger; if wrinkles form on the surface, it is ready.

6 Skim off any scum from the surface, then leave the marmalade to stand for 30 minutes. Stir, then pour into warm, sterilized jars and cover with waxed discs. Leave to cool before adding the lids and labelling the jars.

Citrus Curds

If you have only eaten bought lemon curd, you'll be amazed at the superior flavour of the home-made variety.

Lemon curd

Creamy, buttery and tangy, lemon curd is well worth making yourself. It will taste nothing like the bright yellow bought version, which normally contains little or no egg or butter. Use lemon curd to fill a sponge cake, flavour a trifle or stir it through vanilla ice cream for a quick sweet sensation. Best of all, spoon it generously on to a thick slice of fresh white bread, spread with a little unsalted (sweet) butter.

MAKES ABOUT 450G/1LB

INGREDIENTS

- 3 unwaxed lemons
- 115g/4oz/½ cup caster (superfine) sugar
- 15ml/1 tbsp cornflour (cornstarch), mixed to a paste with 15ml/1 tbsp water
- 2 egg yolks
- 50g/2oz/¼ cup unsalted (sweet) butter

1 Grate the lemon and squeeze the juice. Add the rind and juice to a pan with the sugar and stir over a low heat to dissolve the sugar. Stir in the cornflour paste.

2 Remove from the heat and whisk in the egg yolks. Return to a low heat, whisk for about 2 minutes; remove from the heat. Gradually whisk in the butter. Pour into a sterilized jar, cover and seal at once. Leave to cool, then store in the refrigerator and use within three weeks.

Lemon and lime curd

The tart lime rind and juice gives this spread an added dimension. It makes a lovely filling for a roulade.

MAKES ABOUT 800G/1¾LB

INGREDIENTS

- 3 large unwaxed lemons
- 3 unwaxed limes
- 175g/6oz/¾ lb unsalted (sweet) butter
- 225g/8oz/1 cup caster (superfine) sugar
- 4 large (US extra large) eggs, beaten

1 Wash the lemons and limes and finely grate the rinds.

2 Squeeze the juice from the lemons and limes, then strain it into a bowl.

3 Melt the butter in a heatproof bowl set over a pan of barely simmering water or in the top of a double boiler. Whisk in the rind and juice, the sugar and eggs. Cook, whisking constantly, until the mixture is smooth and thickened. Do not let it boil. Pour into warm, sterilized jars and seal. When cold, store in the refrigerator for up to three weeks.

Lemon and passion fruit curd

This unusual curd tastes wonderful when stirred into thick yogurt or spooned over ice cream. It also makes a great filling for tarts.

MAKES ABOUT 450G/1LB

INGREDIENTS

- 2 passion fruit
- 1 lemon
- 175g/6oz/¾ cup unsalted (sweet) butter
- 115g/4oz/½ cup caster (superfine) sugar
- 2 egg yolks

1 Cut the passion fruit in half and strain all the pulp and seeds into a bowl. Cut the lemon in half and squeeze the juice. Strain it into the bowl.

2 Melt the butter in the top of a double boiler or in a heatproof bowl set over a pan of barely simmering water. Beat in the passion fruit pulp, lemon juice, sugar and egg yolks. Cook, beating constantly, until the mixture thickens. Do not let it boil. Pour into a warm sterilized jar and seal. Leave to cool, then store in the refrigerator and use within three weeks.

Tip for successful citrus curds

The most important thing to remember when making citrus curd is to stir it constantly and avoid letting it boil so that the mixture does not curdle. If you are not confident about making it in a heavy pan, use a double boiler.

Preserving Citrus Fruit

The tradition of preserving citrus fruit goes back to the time when oranges and lemons were a rare luxury. Today there is a plentiful supply of fruit all year round, but it is still worth making preserves, if only to experience the wonderful depth of flavour that develops when orange peel is candied, lemons are preserved in salt or Seville (Temple) oranges metamorphose into marmalade. The best time for preserving citrus fruit is during the winter months when the fruit is at its juiciest.

Quick candied lemon slices

Translucent candied lemon slices look gorgeous on a fresh lemon tart. Use this method when you are going to serve the slices soon after candying.

1 Cut two large, thin-skinned, unwaxed lemons into thin slices, put them in a pan and cover with plenty of water. Simmer for 15–20 minutes, or until the skin is tender, then drain.

2 Put 200g/7oz/scant 1 cup caster (superfine) sugar in a large, shallow pan and stir in about 105ml/7 tbsp water. Heat until the sugar has all dissolved, stirring constantly, then bring to the boil. Boil for 2 minutes, add the lemon slices and cook for 10–15 minutes, until they look shiny and candied.

3 Lift the candied lemon slices out of the syrup using tongs or a slotted spoon and arrange them on top of a tart or cake. Return the syrup to the heat and boil, without stirring, until reduced to a thick glaze. Brush this over the lemon slices and leave to cool completely before serving.

Candying orange peel

When you squeeze oranges for juice, the peel is wasted. Transforming it into candied peel will provide you with a flavoursome ingredient for adding to fruit cakes, scones and desserts. When the candied peel is cool, store it in a clean jar and it will keep for months.

1 Choose unwaxed oranges with thick skins. Wash and dry them. Remove the rind in long thin strips. Using a sharp knife, cut the peel into julienne strips.

2 For each orange you use, put 250ml/8fl oz/1 cup water and 115g/4oz/½ cup granulated sugar into a pan. Stir well over a medium heat until the sugar has dissolved. Bring to the boil, add the strips of orange rind, half-cover the pan and simmer until the syrup has reduced by three-quarters. Remove the pan from the heat and leave to cool completely.

3 Preheat the oven to 110°C/225°F/Gas ¼. Sift icing (confectioners') sugar in a thick, even layer over a baking tray. Drop in the candied peel and roll the pieces in the sugar. Dry in the oven for 1–2 hours, checking often. When the strips are crisp, they are done.

Preserved lemons

These are surprisingly mild and sweet, despite the salt used to preserve them. Although they will be ready after a month, they'll keep for at least another six. They are used all over the Middle East and in North Africa, particularly Morocco, where they form an essential part of the flavouring for a traditional *tagine*. Add strips of preserved lemon to any lamb or chicken stew and you'll notice the difference. The following makes about 1.3kg/3lb.

1 Put 14 unwaxed lemons of average size in a large bowl and pour over cold water to cover. Cut 12 of the lemons into quarters, but be careful to keep them attached at one end.

2 Ease slightly apart the sections on one of the lemons, then spoon 5ml/1 tsp of sea salt into the centre. Close the lemon again. Repeat with the remaining fruit.

3 Pack the salted lemons in sterilized jars, sprinkling with salt, and pressing them down. Squeeze the juice of the remaining 2 lemons over the top, then top up with olive oil. Cover and seal, then store in a cool, dark place.

Citrus Butters and Dressings

Adding flavour to your food is easy when you top a steak, a succulent piece of fish or chicken, or lamb chops with a pat of tangy citrus butter. Lemon vinaigrette gives salads a delicate flavour, quite unlike the usual version using vinegar, and lemon mayonnaise adds a touch of luxury to salads or sandwiches.

Maître d'hôtel butter

This is the classic parsley and lemon butter, frequently used for fish, but also great on steak or chops.

MAKES ABOUT 115G/4OZ/½ CUP

INGREDIENTS

- 115g/4oz/½ cup unsalted (sweet) butter, at room temperature
- 30ml/2 tbsp finely chopped fresh parsley
- 2.5ml/½ tsp lemon juice
- cayenne pepper
- salt and ground black pepper

1 Put the butter into a mixing bowl and soften it by beating it with a spoon. Beat in the parsley, lemon juice and cayenne pepper. Season with salt and pepper.

2 Spread the butter 1cm/¼in thick on a piece of baking parchment, greaseproof (waxed) paper or foil. Chill until firm, then stamp out small rounds or other shapes using pastry (cookie) cutters.

3 If you want to make a butter roll, which can later be cut into slices, spoon the butter on to the foil or paper, then shape it, using the foil or paper as a guide. Do this as quickly as you can, as the heat from your hands will melt the butter. Wrap and chill.

Lemon vinaigrette

The perfect salad dressing for green leaves, lemon vinaigrette is often also called French dressing (except in France, that is). It is basically a mixture of oil and lemon juice, with extra ingredients that vary, depending on what kind of salad it is to be used for. It is a variation on the original vinaigrette, which uses vinegar rather than lemon juice, but lemon has a much more subtle flavour than vinegar. You can make the vinaigrette hours before it is required – just give it a quick whisk to bring it back up to scratch – but don't let green leaves sit in it for too long or they'll wilt. Some cooks mix vinaigrette by the jam-jar-full and keep it in the refrigerator, but the dressing tastes infinitely better when it is freshly made.

MAKES ABOUT 150ML/¼ PINT/⅔ CUP

INGREDIENTS

- 1 garlic clove
- 5ml/1 tsp Dijon mustard
- 45–60ml/3–4 tbsp lemon juice
- 150ml/¼ pint/⅔ cup good quality olive oil
- salt and ground black pepper

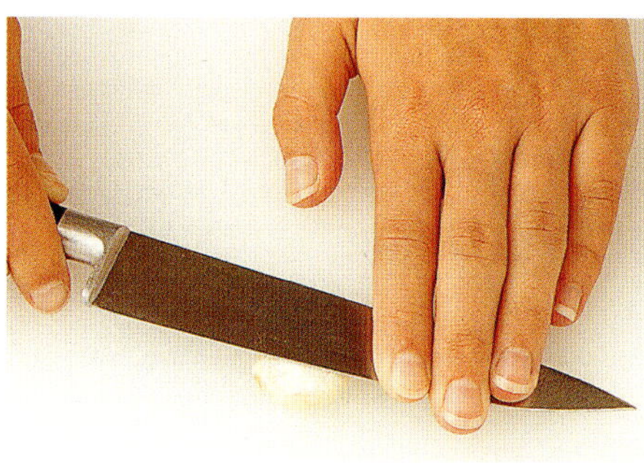

1 Put the garlic on a board, sprinkle it with a little salt and crush with the flat side of the blade of a cook's knife.

2 Scrape the crushed garlic into a small bowl and beat in the mustard, and then the lemon juice.

3 Using a whisk (the horseshoe-shaped type works best), gradually whisk in the oil until the dressing thickens and emulsifies. Season to taste with salt and ground black pepper.

Never-fail mayonnaise

Mayonnaise has many different uses, in sandwiches and hamburgers, as a salad dressing and as a sauce with fish and chicken. This version is made with a food processor. Have all the ingredients at room temperature before you start.

MAKES ABOUT 350ML/12FL OZ/1½ CUPS

INGREDIENTS

- 1 egg, plus 1 egg yolk
- 5ml/1 tsp Dijon mustard
- juice of 1 large lemon
- 175ml/6fl oz/¾ cup olive oil
- 175ml/6fl oz/¾ cup rapeseed (canola), sunflower or corn oil
- salt and white pepper

1 Process the whole egg and yolk in a food processor for 20 seconds. Add the mustard, half the lemon juice, and a pinch of salt and pepper. Process for about 30 seconds until mixed.

2 With the motor running, pour in the oils through the feeder tube in a thin stream. Process until the mayonnaise is thick. Taste and add more lemon juice and seasoning if necessary.

Simple mayonnaise variations

- **Aioli** Add 4 garlic cloves with the eggs when making mayonnaise.
- **Remoulade** To 300ml/½ pint/ 1¼ cups mayonnaise, add 15ml/ 1 tbsp each of chopped fresh chives, tarragon and parsley, Dijon mustard and chopped gherkins, and 5ml/1 tsp each of anchovy essence and chopped capers.

Citrus Sauces

One of the many qualities that makes citrus fruit so special is its affinity for both sweet and savoury dishes.

Hollandaise sauce

This classic sauce has a reputation for being difficult to make, as it separates if it gets too hot. This version is much easier. Serve it with asparagus, artichokes, poached eggs, salmon and fish mousses.

MAKES ABOUT 300ML/½ PINT/1¼ CUPS

INGREDIENTS
- 3 egg yolks
- 15–30m/1–2 tbsp lemon juice
- 175g/6oz/¾ cup unsalted (sweet) butter, diced
- salt and ground black pepper

1 Add the egg yolks and 15ml/1 tbsp of the lemon juice to a heavy pan. Season and whisk well to combine.

2 Add the butter and set the pan over a medium heat. Whisk constantly so that as the butter melts, it is blended into the egg yolks. Continue whisking until thick.

3 Taste the sauce and add more lemon juice, salt and pepper if needed.

Sauce bigarade

The classic orange sauce to serve with duck and rich game.

MAKES ABOUT 550ML/18FL OZ/2½ CUPS

INGREDIENTS
- roasting pan juices or 25g/1oz/ 2 tbsp butter
- 40g/1½oz/3 tbsp plain (all-purpose) flour
- 300ml/½ pint/1¼ cups hot stock
- 150ml/¼ pint/⅔ cup red wine
- 2 unwaxed Seville (Temple) oranges or 2 sweet oranges plus 10ml/2 tsp lemon juice
- 15ml/1 tbsp orange liqueur
- 30ml/2 tbsp redcurrant jelly
- salt and ground black pepper

1 Pour off any excess fat from the roasting pan, leaving the juices, or melt the butter in a small pan. Add the flour and cook, stirring constantly, for about 4 minutes, or until lightly browned.

2 Remove from the heat and gradually blend in the stock and wine. Bring to the boil, stirring constantly. Lower the heat and simmer gently for 5 minutes.

3 Using a zester, pare the rind from one orange. Squeeze the juice from both fruit.

4 Blanch the rind by putting it in a pan, pouring over boiling water to cover and cooking for 5 minutes. Drain the rind, then return it to the pan. Add the sauce.

5 Add the orange juice and lemon juice, if using, to the pan with the liqueur and jelly, stirring to dissolve the jelly. Season to taste and serve.

Lemon custard

In sweet sauces, citrus fruit really stars. This creamy sauce is perfect spooned over a steamed lemon sponge pudding, or poured over the sponges in a trifle. It is also delicious with sliced bananas or chilled and served just as it is with dessert cookies.

MAKES ABOUT 600ML/1 PINT/2½ CUPS

INGREDIENTS
- 15ml/1 tbsp cornflour (cornstarch)
- 15ml/1 tbsp custard powder (see Cook's tip)
- 15ml/1 tbsp caster (superfine) sugar
- finely grated rind of 2 lemons
- 600ml/1 pint/⅔ cup milk
- 15g/½oz/1 tbsp butter

1 Mix the cornflour and custard powder in a large bowl. Stir in the sugar and finely grated lemon rind.

2 Stir in 15–30ml/1–2 tbsp of the milk and mix to a smooth paste. Heat the remaining milk in a heavy pan. As soon as bubbles start to appear on the surface, around the edge, remove the pan from the heat.

3 Gradually stir the hot milk into the paste, until the mixture is very smooth. Return it to the pan and heat gently, stirring until thick. Do not allow the custard to scorch or to boil vigorously. Whisk in the butter to finish.

COOK'S TIP
You may substitute cornflour (cornstarch) instead. The colour will not be so bright, but the flavour will not be compromised.

Soups and Appetizers

Summery soups and tangy appetizers are an exciting and effective way to get the taste buds tingling at the start of a meal. Refreshing citrus flavours, such as the piquancy of fresh and preserved lemons or the rich sweetness of oranges, cleanse the palate and stimulate the appetite. Recipes range from creamy and fragrant soups to exotically spiced chicken and subtly flavoured vegetable dishes, all of them offering the special lift and zest that comes from these most versatile of fruits.

AVGOLEMONO

THIS IS THE MOST POPULAR OF GREEK SOUPS. THE NAME MEANS EGG AND LEMON, THE TWO MOST IMPORTANT INGREDIENTS, WHICH PRODUCE A LIGHT, NOURISHING SOUP. ORZO IS GREEK, RICE-SHAPED PASTA, BUT YOU CAN USE ANY SMALL SOUP PASTA.

SERVES FOUR TO SIX

INGREDIENTS
- 1.75 litres/3 pints/7½ cups chicken stock
- 115g/4oz/½ cup orzo pasta
- 3 eggs
- juice of 1 large lemon
- salt and ground black pepper
- lemon slices, to garnish

COOK'S TIP

This egg and lemon combination is also widely used in Greece as a sauce for pasta or with meatballs.

1 Pour the stock into a large pan, and bring it to a rolling boil. Add the pasta and cook for 5 minutes.

2 Beat the eggs until frothy, then add the lemon juice and 15ml/1 tbsp of cold water. Slowly stir in a ladleful of the hot chicken stock, then add one or two more. Return this mixture to the pan, remove from the heat and stir well.

3 Season to taste with salt and pepper and serve immediately, garnished with lemon slices. (Do not let the soup boil once the eggs have been added or it will curdle.)

CARROT AND ORANGE SOUP

This traditional, bright and summery soup is always popular for its wonderfully creamy consistency and vibrantly fresh citrus flavour. Use a good, home-made chicken or vegetable stock if you can for the best results.

SERVES FOUR

INGREDIENTS

- 50g/2oz/¼ cup butter
- 3 leeks, sliced
- 450g/1lb carrots, sliced
- 1.2 litres/2 pints/5 cups chicken or vegetable stock
- rind and juice of 2 oranges
- 2.5ml/½ tsp freshly grated nutmeg
- 150ml/¼ pint/⅔ cup Greek (US strained plain) yogurt
- salt and ground black pepper
- fresh sprigs of coriander (cilantro), to garnish

1 Melt the butter in a large pan. Add the leeks and carrots and stir well, coating the vegetables with the butter. Cover and cook for about 10 minutes, until the vegetables are beginning to soften but not colour.

2 Pour in the stock and the orange rind and juice. Add the nutmeg and season to taste with salt and pepper. Bring to the boil, lower the heat, cover and simmer for about 40 minutes, or until the vegetables are tender.

3 Leave to cool slightly, then purée the soup in a food processor or blender until smooth.

4 Return the soup to the pan and add 30ml/2 tbsp of the yogurt, then taste the soup and adjust the seasoning, if necessary. Reheat gently.

5 Ladle the soup into warm individual bowls and put a swirl of yogurt in the centre of each. Sprinkle the fresh sprigs of coriander over each bowl to garnish, and serve immediately.

GLOBE ARTICHOKES WITH BEANS AND AIOLI

MAKE THE MOST OF FRESH ARTICHOKES WHEN IN SEASON BY SERVING THEM WITH A SQUEEZE OF LEMON AND THIS DELICIOUS GARLIC AND MAYONNAISE DRESSING FROM SPAIN.

SERVES THREE

INGREDIENTS
- 225g/8oz green beans
- 3 small globe artichokes
- 15ml/1 tbsp olive oil
- pared rind of 1 lemon
- coarse salt, for sprinkling
- lemon wedges, to garnish

For the aioli
- 6 large garlic cloves, thinly sliced
- 10ml/2 tsp white wine vinegar
- 250ml/8fl oz/1 cup olive oil
- salt and ground black pepper

1 First, make the aioli. Put the garlic and vinegar in a food processor or blender. With the motor running, slowly pour in the olive oil through the lid or feeder tube until the mixture is quite thick and smooth. (Alternatively, crush the garlic to a paste with the vinegar and gradually beat in the oil using a hand whisk.) Season with salt and pepper to taste.

2 To make the salad, cook the green beans in lightly salted boiling water for 1–2 minutes until slightly softened. Drain well.

3 Trim the artichoke stalks close to the base. Cook the artichokes in a large pan of salted water for about 30 minutes, or until you can easily pull away a leaf from the base. Drain well.

4 Using a large, sharp knife, cut the artichokes in half lengthways and carefully scrape out the hairy choke using a teaspoon.

5 Arrange the artichokes and beans on serving plates and drizzle with the olive oil. Sprinkle the lemon rind over them and season to taste with coarse salt and a little pepper. Spoon the aioli into the artichoke hearts and serve the salad warm, garnished with lemon wedges.

6 To eat the artichokes, squeeze a little lemon juice over them, then pull the leaves from the base one at a time and use to scoop a little of the aioli sauce. Gently scrape away the fleshy end of each leaf with your teeth and discard the remainder of the leaf. Eat the tender base or "heart" of the artichoke with a knife and fork.

OLIVES WITH MOROCCAN LEMON MARINADES

Preserved and fresh lemons provide their own distinct qualities to these appetizers from North Africa. Start preparations a week in advance to allow the flavours to develop.

SERVES SIX TO EIGHT

INGREDIENTS

- 450g/1lb/2⅔ cups olives (unpitted)

For the piquant marinade

- 45ml/3 tbsp chopped fresh coriander (cilantro)
- 45ml/3 tbsp chopped fresh parsley
- 1 garlic clove, finely chopped
- good pinch of cayenne pepper
- good pinch of ground cumin
- 30–45ml/2–3 tbsp olive oil
- 30–45ml/2–3 tbsp lemon juice

For the spicy marinade

- 60ml/4 tbsp chopped fresh coriander (cilantro)
- 60ml/4 tbsp chopped fresh parsley
- 1 garlic clove, finely chopped
- 5ml/1 tsp grated fresh root ginger
- 1 red chilli, seeded and sliced
- ¼ preserved lemon, cut into strips

1 Using the flat side of a large knife blade, crack the olives, hard enough to break the flesh, but taking care not to crack the pit. Place the olives in a bowl of cold water, cover and leave overnight in a cool place to remove the excess brine. Next day, drain the olives thoroughly and divide them among two sterilized screwtop jars.

2 To make the piquant marinade, mix the coriander, parsley and garlic together in a bowl. Add the cayenne pepper and cumin and 30ml/2 tbsp of the olive oil and lemon juice. Add the olives from one jar, mix well and return to the jar. Add more olive oil and lemon juice to cover if necessary. Seal.

3 To make the spicy marinade, mix together the coriander, parsley, garlic, ginger, chilli and preserved lemon. Add the olives from the second jar, mix well and return to the jar. Seal.

4 Store the olives in the refrigerator for at least one week before using, shaking the jars occasionally.

ASPARAGUS WITH EGG AND LEMON SAUCE

EGGS AND LEMONS ARE OFTEN FOUND IN DISHES FROM GREECE, TURKEY AND THE MIDDLE EAST. THIS SAUCE HAS A FRESH, TANGY TASTE AND BRINGS OUT THE BEST IN ASPARAGUS.

SERVES FOUR

INGREDIENTS

- 675g/1½lb asparagus, tough ends removed, and tied in a bundle
- 15ml/1 tbsp cornflour (cornstarch)
- about 10ml/2 tsp sugar
- 2 egg yolks
- juice of 1½ lemons
- salt

VARIATIONS

This sauce goes very well with young vegetables. Try it with baby leeks, cooked whole or chopped, or serve it with other baby vegetables, such as carrots and courgettes (zucchini).

1 Cook the bundle of asparagus in salted boiling water for 7–10 minutes. Drain well and arrange the asparagus in a serving dish. Reserve 200ml/7fl oz/ scant 1 cup of the cooking liquid.

2 Blend the cornflour with the cooled, reserved cooking liquid and place in a small pan. Bring to the boil, stirring constantly, and cook over a low heat until the sauce thickens slightly. Stir in 10ml/2 tsp sugar, then remove the pan from the heat and leave to cool slightly.

3 Beat the egg yolks thoroughly with the lemon juice and then stir gradually into the cooled sauce. Cook over a very low heat, stirring constantly, until the sauce is fairly thick. Be careful not to overheat the sauce or it may curdle. As soon as the sauce has thickened, remove the pan from the heat and continue stirring for 1 minute. Taste and add salt or sugar as necessary. Leave the sauce to cool slightly.

4 Stir the cooled sauce, then pour a little over the asparagus. Cover and chill for at least 2 hours before serving with the rest of the sauce.

COOK'S TIP

Use tiny asparagus spears for an elegant first course or a special dinner party.

Prawn Cocktail

There is no nicer appetizer than a good, fresh prawn cocktail, dressed in a creamy mayonnaise flavoured with lemon and Worcestershire sauce.

SERVES SIX

INGREDIENTS

- 60ml/4 tbsp double (heavy) cream, lightly whipped
- 60ml/4 tbsp mayonnaise, preferably home-made
- 60ml/4 tbsp tomato ketchup
- 5–10ml/1–2 tsp Worcestershire sauce
- juice of 1 lemon
- ½ cos or romaine lettuce, or other very crisp lettuce
- 450g/1lb cooked peeled prawns (shrimp)
- salt, ground black pepper and paprika
- 6 large whole cooked prawns in the shell, to garnish (optional)
- thinly sliced and buttered brown bread and lemon wedges, to serve

1 Lightly whisk the cream, mayonnaise and tomato ketchup together in a bowl. Add Worcestershire sauce to taste, then whisk in enough of the lemon juice to make a tangy sauce.

COOK'S TIP

Partly peeled prawns, with the tail "fan" intact, make a pretty garnish.

2 Finely shred the lettuce and fill six individual glasses one-third full.

3 Stir the prawns into the sauce, then season with salt and pepper. Spoon the mixture over the lettuce. If you like, drape a whole prawn over the edge of each glass, and sprinkle the cocktails with black pepper and/or paprika.

Chinese Chicken Wings

Lemon is frequently used for marinades in Chinese cuisine. Here it combines with ginger, chilli and garlic to give roasted chicken wings an exotic flavour. The wings make a tasty first course; as they are best eaten with the fingers, make sure you provide finger bowls and plenty of paper napkins.

SERVES FOUR

INGREDIENTS

- 12 chicken wings
- 3 garlic cloves, crushed
- 4cm/1½in piece fresh root ginger, peeled and grated
- juice of 1 large lemon
- 45ml/3 tbsp soy sauce
- 45ml/3 tbsp clear honey
- 2.5ml/½ tsp chilli powder
- 150ml/¼ pint/⅔ cup chicken stock
- salt and ground black pepper
- lemon and lime wedges, to garnish

1 Remove the wing tips (pinions) and discard. Cut the wings into two pieces.

2 Put the chicken pieces into a shallow dish, cover and set aside.

3 Mix together the garlic, ginger, lemon and soy sauce in a small jug (pitcher). Blend in the honey, chilli powder, seasoning and chicken stock. Pour the mixture over the chicken wings and turn to coat completely. Cover and marinate in the refrigerator overnight.

4 Preheat the oven to 220°C/425°F/Gas 7. Remove the chicken wings from the marinade and arrange them in a single layer in a large roasting pan. Roast for about 25 minutes, generously basting at least twice with the marinade during cooking.

5 Place the wings on a serving plate. Add the remaining marinade to the roasting pan and bring to the boil. Cook until it turns a syrupy consistency and spoon a little over the wings. Serve the chicken wings immediately, garnished with the lemon and lime wedges.

Carpaccio with Rocket

Invented in Venice, carpaccio is named in honour of the Renaissance painter. In this fine Italian dish raw beef is lightly dressed with lemon juice and olive oil and is traditionally served with flakes of Parmesan cheese. Use very fresh meat of the best quality and ask the butcher to slice it very thinly.

SERVES FOUR

INGREDIENTS

- 1 garlic clove, peeled and cut in half
- 1½ lemons
- 50ml/2fl oz/¼ cup extra virgin olive oil
- 2 bunches rocket (arugula)
- 4 very thin slices of beef fillet
- 115g/4oz Parmesan cheese, shaved
- salt and ground black pepper

1 Rub the cut side of the garlic over the inside of a bowl. Squeeze the lemons into the bowl, then whisk in the olive oil. Season with salt and pepper. Leave to stand for at least 15 minutes.

2 Carefully wash the rocket and tear off any thick stalks. Spin or pat dry with kitchen paper. Arrange the rocket around the edge of a serving platter or divide among four individual plates.

3 Place the sliced beef in the centre of the platter, and pour the sauce over it, spreading it evenly over the meat. Arrange the shaved Parmesan on top of the meat slices and serve immediately.

VARIATION

You can also serve meaty fish, such as tuna, in the same way. Place a tuna steak between sheets of clear film (plastic wrap) and pound with a rolling pin. Then roll it up tightly and wrap in clear film. Place in the freezer for about 4 hours until firm. Unwrap and, using a very sharp knife, cut the fish crossways into very thin slices. Serve with the rocket (arugula), sauce and Parmesan.

Vegetables and Vegetarian Dishes

Citrus flavours pep up even the most mundane of vegetables, and give salads a refreshing lift. Besides such classic combinations as carrots and oranges, the recipes feature some more unusual and utterly delicious mixtures, from orange-glazed potatoes to spicy cauliflower with a hint of lemon. Used in combination with herbs and spices, oranges and lemons also breathe new life into vegetarian pasta and rice dishes, as well as playing an important nutritional role: the presence of vitamin C is what enables the body to absorb iron from vegetable sources.

Peas with Baby Onions and Cream

A creamy sauce with a hint of lemon is the perfect choice for sweet baby onions and fresh garden peas. This makes a delightful, summery side dish that is versatile enough to accompany grilled meat, chicken or fish.

2 Add the peas and stir-fry briefly. Add 120ml/4fl oz/½ cup water, bring to the boil and simmer for about 10 minutes until tender. There should be a thin layer of water on the base of the pan.

3 Blend the cream with the flour. Remove the pan from the heat and stir in the cream mixture and parsley. Season to taste with salt and pepper.

SERVES FOUR

INGREDIENTS

- 175g/6oz baby (pearl) onions
- 15g/½oz/1 tbsp butter
- 900g/2lb fresh peas or 350g/12oz/3 cups shelled or frozen peas
- 150ml/¼ pint/⅔ cup double (heavy) cream
- 15g/½oz/2 tbsp plain (all-purpose) flour
- 10ml/2 tsp chopped fresh parsley
- 15–30ml/1–2 tbsp lemon juice
- salt and ground black pepper

1 Remove the skin from the onions and halve any large ones. Melt the butter in a frying pan and cook them for about 5 minutes until tender and just golden.

4 Cook over a low heat for 3 minutes, or until the sauce has thickened slightly. Add lemon juice to taste, stir through, and serve immediately.

Hasselback Potatoes

This is an unusual way to cook with potatoes: each potato half is sliced almost to the base and then roasted with oil and butter. The crispy potatoes are then coated in an orange glaze and returned to the oven until deep golden brown and crunchy.

SERVES FOUR

INGREDIENTS

- 4 large potatoes
- 25g/1oz/2 tbsp butter, melted
- 45ml/3 tbsp olive oil

For the glaze

- juice of 1 orange
- grated rind of ½ orange
- 15ml/1 tbsp demerara (raw) sugar
- ground black pepper

COOK'S TIP

To make potato fans, cut through the potato halves at a slight angle, without cutting all the way through. Gently press on the top of the potato until it flattens out and fans slightly.

1 Preheat the oven to 190°C/375°F/ Gas 5. Peel the potatoes and cut in half lengthways. Place the flat side down on the chopping board and then cut down as if making very thin slices, but leaving the bottom 1cm/½in intact.

2 Place the potatoes in a large roasting pan. Coat them with the melted butter and pour the olive oil over and around. Roast the potatoes for 50 minutes.

3 Meanwhile, make the glaze. Place the orange juice, orange rind and sugar in a pan and heat gently, stirring, until the sugar has dissolved. Simmer over a low heat for 3–4 minutes until thick, then remove the pan from the heat.

4 When the potatoes begin to brown, brush them all over with the orange glaze and return them to the oven for a further 15 minutes, or until golden brown. Tip on to a warmed serving plate and serve immediately.

Spiced Turnip, Spinach and Tomato Stew

Paprika and lemon juice add a delicate piquancy to sweet baby turnips, tender spinach and ripe tomatoes in this simple Eastern Mediterranean vegetable stew. Use plum tomatoes or vine-ripened tomatoes for the best and most intense flavour.

SERVES SIX

INGREDIENTS

- 450g/1lb plum or other well-flavoured tomatoes
- 60ml/4 tbsp olive oil
- 2 onions, sliced
- 450g/1lb baby turnips, peeled
- 5ml/1 tsp paprika
- 2.5ml/½ tsp caster (superfine) sugar
- 60ml/4 tbsp chopped fresh coriander (cilantro)
- 450g/1lb fresh young spinach, stalks removed
- 15–30ml/1–2 tbsp lemon juice
- salt and ground black pepper

1 Make a small cross in the tops of the tomatoes, plunge them into a bowl of boiling water for 30 seconds, then refresh in a bowl of cold water. Peel off the skins and chop the tomatoes.

2 Heat the olive oil in a large, heavy frying pan or sauté pan and cook the onion slices for about 5 minutes, or until golden brown.

3 Add the baby turnips, tomatoes and paprika to the pan with 60ml/4 tbsp water and cook over a medium heat until the tomatoes are pulpy. Cover with a lid and continue cooking until the baby turnips have softened.

4 Stir in the sugar and coriander, then add the spinach and salt and pepper. Add lemon juice to taste and cook for a further 2–3 minutes, or until the spinach has wilted. Serve warm or cold.

Lemon-Spiced Red Cabbage

Pears and red cabbage make delicious partners, and here their flavours are enhanced with red wine, lemon and caraway. Serve this East European vegetable dish as an accompaniment to roast pork or game, such as wild boar or venison.

SERVES SIX TO EIGHT

INGREDIENTS

- 3 thick rindless bacon rashers (strips), diced
- 1 large onion, chopped
- 1 large red cabbage, evenly shredded
- 3 garlic cloves, crushed
- 15–25ml/1–1½ tbsp caraway seeds
- 120ml/4fl oz/½ cup water
- 2 firm, ripe pears, cored and evenly chopped
- juice of 1 lemon
- 475ml/16fl oz/2 cups red wine
- 45ml/3 tbsp red wine vinegar
- 150g/5oz/⅔ cup clear honey
- salt and ground black pepper
- caraway seeds and chopped fresh chives, to garnish

1 Dry-fry the diced bacon in a pan over a gentle heat for 5–10 minutes, or until golden brown.

2 Add the onion and cook, stirring occasionally, for 5 minutes, or until pale golden brown.

3 Stir the cabbage, garlic, caraway seeds and the water into the pan. Cover and cook for 8–10 minutes.

4 Season to taste with salt and pepper, then add the chopped pears, lemon juice, red wine and wine vinegar. Cover and cook over a low heat for about 15 minutes. Stir in the honey.

5 If there is too much cooking liquid, remove the lid and continue to cook to allow it to reduce. The pears will have broken up in the pan, and the quantity will have reduced by one-third. Adjust the seasoning to taste, if necessary. Transfer to a warm serving dish and serve sprinkled with caraway seeds and chopped fresh chives.

Cauliflower with Tomatoes and Cumin

Warm spices are sharpened with lemon juice in this unusual side dish of cauliflower braised with tomatoes. It is excellent served with barbecue-cooked meat or fish.

SERVES FOUR

INGREDIENTS
- 4 tomatoes
- 30ml/2 tbsp sunflower or olive oil
- 1 onion, chopped
- 1 garlic clove, crushed
- 1 small cauliflower, broken into florets
- 5ml/1 tsp cumin seeds
- a good pinch of ground ginger
- 15–30ml/1–2 tbsp lemon juice
- 30ml/2 tbsp chopped fresh coriander (cilantro) (optional)
- salt and ground black pepper

1 Cut a small cross in the top of the tomatoes, plunge them into boiling water for 30 seconds, then refresh in cold water. Peel off the skins, cut the tomatoes into quarters and remove and discard the seeds.

2 Heat the oil in a flameproof casserole, add the onion and garlic and stir-fry for 2–3 minutes, or until the onion is softened. Add the cauliflower and stir-fry for a further 2–3 minutes, or until the cauliflower is flecked with brown. Add the cumin seeds and ginger, fry briskly for 1 minute, and then add the tomatoes, 175ml/6fl oz/¾ cup water and some salt and pepper.

3 Bring to the boil and then reduce the heat, cover with a plate or with foil and simmer for 6–7 minutes, or until the cauliflower is just tender.

4 Stir in a little lemon juice to sharpen the flavour, and taste and adjust the seasoning if necessary. Sprinkle over the chopped coriander, if using, and serve immediately.

Roasted Mediterranean Vegetables

Oven roasting brings out the robust flavours of courgette, onion, pepper and tomato. Toss them in a piquant lemon and harissa dressing for extra tang.

SERVES FOUR

INGREDIENTS
- 2–3 courgettes (zucchini)
- 1 Spanish onion
- 2 red (bell) peppers
- 16 cherry tomatoes
- 2 garlic cloves, chopped
- pinch of cumin seeds
- 5ml/1 tsp fresh thyme or 4–5 torn basil leaves
- 60ml/4 tbsp olive oil
- juice of ½ lemon
- 5–10ml/1–2 tsp harissa or Tabasco sauce
- fresh thyme sprigs, to garnish

COOK'S TIP
Harissa is a fiery-hot sauce from Tunisia. It is usually sold in jars and can be found in Middle Eastern stores.

1 Preheat the oven to 220°C/425°F/Gas 7. Trim the courgettes and cut into long strips. Cut the onion into thin wedges. Cut the peppers into chunks, discarding the seeds and core.

2 Place the vegetables in a roasting pan, add the tomatoes, chopped garlic, cumin seeds and thyme or basil.

3 Sprinkle with the olive oil and toss to coat. Roast for 25–30 minutes, or until the vegetables are very soft and slightly charred at the edges.

4 Blend the lemon juice with the harissa or Tabasco sauce and stir into the vegetables before serving, garnished with the thyme.

Carrot and Orange Salad

This is a wonderful, fresh-tasting salad with such a fabulous combination of citrus fruit and vegetables that it is difficult to know whether it is a salad or a dessert. It makes a refreshing accompaniment to grilled meat, chicken or fish.

SERVES FOUR

INGREDIENTS
- 450g/1lb carrots
- 2 large Navelina oranges, such as Washington or Bahia
- 15ml/1 tbsp extra virgin olive oil
- 30ml/2 tbsp freshly squeezed lemon juice
- pinch of sugar (optional)
- 30ml/2 tbsp chopped pistachio nuts or toasted pine nuts
- salt and ground black pepper

1 Peel the carrots and coarsely grate them into a large bowl.

2 Cut a thin slice of peel and pith from each end of the oranges. Place cut-side down on a plate and cut off the peel and pith in strips. Holding the oranges over a bowl, cut out each segment leaving the membrane behind. Squeeze the juice from the membrane.

3 Blend the oil, lemon juice and orange juice in a small bowl. Season with salt and pepper, and sugar, if you like.

4 Toss the oranges with the carrots and pour the dressing over. Sprinkle over the pistachios or pine nuts and serve.

Chargrilled Pepper Salad

The simple addition of fresh basil and coriander to this pepper and pasta salad give it loads of flavour. A lemon and pesto dressing brings all the flavours together making this a superb summer side salad that would be ideal for an al fresco meal.

SERVES FOUR

INGREDIENTS
- 1 large red (bell) pepper
- 1 large green (bell) pepper
- 250g/9oz/2¼ cups dried fusilli tricolore
- 1 handful fresh basil leaves
- 1 handful fresh coriander (cilantro) leaves
- 1 garlic clove
- salt and ground black pepper

For the dressing
- 30ml/2 tbsp pesto
- juice of ½ lemon
- 60ml/4 tbsp extra virgin olive oil

1 Put the peppers under a hot grill (broiler) and grill (broil) them for about 10 minutes, turning frequently until they are charred on all sides. Put the hot peppers in a plastic bag, seal the bag and set aside until the peppers are cool.

2 Meanwhile, bring a large pan of salted water to the boil. Add the pasta and cook according to the instructions on the packet.

3 Whisk all the dressing ingredients together in a large mixing bowl. Drain the cooked pasta and tip it into the bowl of dressing. Toss well to mix and set aside to cool.

COOK'S TIP
Fusilli tricolore are small, red, white and green pasta spirals.

4 Rinse the cooled peppers under cold running water. Peel off the skins with your fingers, split the peppers open and pull out the cores. Rub off all the seeds under the running water, then pat dry on kitchen paper.

5 Chop the peppers and add them to the pasta. Put the basil, coriander and garlic on a board and chop them. Add to the pasta and toss well to mix, then taste and adjust the seasoning, if necessary, and serve.

Peppers Filled with Spiced Vegetables

Indian spices and lemon season the potato and aubergine stuffing in these colourful baked peppers. They are excellent with plain rice and a lentil dhal, or they can be served with salad, Indian breads and a cucumber and yogurt raita.

SERVES SIX

INGREDIENTS

- 6 large evenly shaped red or yellow (bell) peppers
- 500g/1¼lb waxy potatoes
- 1 small onion, chopped
- 4–5 garlic cloves, chopped
- 5cm/2in piece fresh root ginger, peeled and chopped
- 1–2 fresh green chillies, seeded and chopped
- 105ml/7 tbsp water
- 90ml/6 tbsp groundnut (peanut) oil
- 1 aubergine (eggplant), cut into 1cm/½in dice
- 10ml/2 tsp cumin seeds
- 5ml/1 tsp kalonji seeds (see Cook's Tip)
- 2.5ml/½ tsp ground turmeric
- 5ml/1 tsp ground coriander
- 5ml/1 tsp ground toasted cumin seeds
- pinch or two of cayenne pepper
- about 45ml/3 tbsp lemon juice
- salt and ground black pepper
- 30ml/2 tbsp chopped fresh coriander (cilantro), to garnish

1 Cut the tops off the peppers then remove and discard the seeds. Cut a thin slice off the base, if necessary, to make them stand upright.

2 Bring a large pan of lightly salted water to the boil. Add the peppers and cook for 5–6 minutes. Drain and leave the peppers upside down in a colander.

3 Cook the potatoes in salted boiling water for 10–12 minutes, or until just tender. Drain, cool and peel, then cut into 1cm/½in dice.

4 Put the onion, garlic, ginger and green chillies in a food processor or blender with about 60ml/4 tbsp of the measured water and process to a purée.

5 Heat 45ml/3 tbsp of the oil in a large frying pan, add the aubergine and cook, stirring occasionally, until browned on all sides. Remove from the pan and set aside. Add another 30ml/2 tbsp of the oil to the pan and cook the potatoes until lightly browned. Remove from the pan and set aside.

6 If necessary, add another 15ml/1 tbsp oil to the pan. Add the cumin and kalonji seeds. Cook briefly until the seeds darken, then add the turmeric, coriander and ground cumin. Cook for 15 seconds. Stir in the onion and garlic purée and cook, scraping the pan with a spatula, until it begins to brown.

7 Return the potatoes and aubergines to the pan, season with salt, pepper and a pinch or two of cayenne. Add the remaining water and 30ml/2 tbsp lemon juice and then cook, stirring, until the liquid evaporates. Preheat the oven to 190°C/375°F/Gas 5.

8 Fill the peppers with the potato mixture and place on a lightly greased baking sheet. Brush the peppers with a little oil and bake for 30–35 minutes, until the peppers are cooked through and tender. Leave to cool slightly, then sprinkle with a little more lemon juice, garnish with the chopped coriander and serve immediately.

COOK'S TIP

Kalonji, or nigella as it is sometimes known, is a tiny black seed. It is widely used in Indian cooking, often sprinkled over breads or in potato dishes. It has a mild, slightly nutty flavour and is best toasted for a few seconds in a dry frying pan over a medium heat. This helps to bring out its flavour.

Lemony Okra and Tomato Tagine

In this spicy vegetable dish, the heat of the chilli is offset by the refreshing flavour of lemon juice. This is a surprisingly substantial dish, and ideal for a vegetarian meal.

SERVES FOUR

INGREDIENTS

- 350g/12oz okra
- 5–6 tomatoes
- 2 small onions
- 2 garlic cloves, crushed
- 1 fresh green chilli, seeded
- 5ml/1 tsp paprika
- small handful of fresh coriander (cilantro), plus extra to garnish
- 30ml/2 tbsp sunflower oil
- juice of 1 lemon

1 Trim the okra and then cut them into 1cm/½in lengths. Set aside.

2 Plunge the tomatoes into a bowl of boiling water for 30 seconds, then drain and refresh in cold water. Peel off the skins and coarsely chop the flesh. Set the tomatoes aside.

3 Coarsely chop one of the onions and place it in a food processor or blender with the garlic, green chilli, paprika, fresh coriander and 60ml/4 tbsp water. Process to a paste.

4 Thinly slice the second onion and cook in the oil for 5–6 minutes, or until golden brown. Transfer to a plate with a slotted spoon.

5 Reduce the heat and pour in the onion and coriander mixture. Cook for 1–2 minutes, stirring frequently, and then add the okra, tomatoes, lemon juice and about 120ml/4fl oz/½ cup water. Stir well to mix, cover tightly and simmer gently over a low heat for about 15 minutes, or until the okra is tender.

6 Transfer to a warmed serving dish, sprinkle with the fried onion rings, garnish with coriander and serve immediately.

COOK'S TIP

Okra is also known as bhindi, gumbo and ladies' fingers.

Leek, Mushroom and Lemon Risotto

THE DELICIOUS COMBINATION OF LEEKS AND LEMON IS PERFECT IN THIS LIGHT RISOTTO; BROWN CAP MUSHROOMS PROVIDE ADDITIONAL TEXTURE AND EXTRA FLAVOUR.

SERVES FOUR

INGREDIENTS

- 225g/8oz trimmed leeks
- 225g/8oz/3 cups brown cap (cremini) mushrooms
- 30ml/2 tbsp olive oil
- 3 garlic cloves, crushed
- 75g/3oz/6 tbsp butter
- 1 large onion, coarsely chopped
- 350g/12oz/1¾ cups risotto rice, such as arborio or carnaroli
- 1.2 litres/2 pints/5 cups simmering vegetable stock
- grated rind of 1 lemon
- 45ml/3 tbsp lemon juice
- 50g/2oz/⅔ cup freshly grated Parmesan cheese
- 60ml/4 tbsp mixed chopped fresh chives and flat leaf parsley
- salt and ground black pepper

1 Slice the leeks in half lengthways, wash them well and then slice them evenly. Wipe the mushrooms with kitchen paper and chop them coarsely.

2 Heat the oil in a large pan and cook the garlic for 1 minute. Add the leeks, mushrooms and plenty of seasoning and cook over a medium heat for about 10 minutes, or until the leeks have softened and browned. Spoon into a bowl and set aside.

3 Add 25g/1oz/2 tbsp of the butter to the pan. When it has melted, add the onion and cook over a medium heat for 5 minutes, or until it has softened and is golden.

4 Stir in the rice and cook for about 1 minute, or until the grains begin to look translucent and are coated in the fat. Add a ladleful of stock and cook gently, stirring occasionally, until the liquid has been absorbed.

5 Continue to add stock, a ladleful at a time, until all of it has been absorbed, stirring constantly. This will take about 30 minutes. The risotto will become thick and creamy, and the rice should be tender, but not sticky.

6 Just before serving, add the leeks and mushrooms with the remaining butter. Stir in the grated lemon rind and juice. Add the grated Parmesan cheese and the herbs. Adjust the seasoning if necessary and serve immediately.

COOK'S TIP

Risotto rice is a rounder grain than long grain and is capable of absorbing a lot of liquid, which gives it a creamy texture.

VERMICELLI WITH LEMON

FRESH AND TANGY, THIS MAKES AN EXCELLENT FIRST COURSE FOR A DINNER PARTY. IT DOES NOT RELY ON FRESH SEASONAL INGREDIENTS, AS LEMONS PROVIDE THE ESSENTIAL FLAVOURING, SO IT IS GOOD AT ANY TIME OF YEAR. IT IS ALSO AN IDEAL CHOICE WHEN YOU ARE PUSHED FOR TIME.

SERVES FOUR

INGREDIENTS

- 350g/12oz dried vermicelli
- juice of 2 large lemons
- 50g/2oz/¼ cup butter
- 200ml/7fl oz/scant 1 cup single (light) cream
- 115g/4oz/1⅓ cups freshly grated Parmesan cheese
- salt and ground black pepper

1 Bring a large pan of lightly salted water to the boil. Add the pasta, bring back to the boil and cook according to the instructions on the packet.

COOK'S TIP

Lemons vary in the amount of juice they yield and are best squeezed at room temperature. On average, a large fresh lemon will yield 60–90ml/4–6 tbsp. The flavour of this dish is quite sharp – you can use less juice if you prefer.

2 While the pasta is cooking, pour the lemon juice into a medium pan. Add the butter and cream, then season with salt and pepper to taste.

3 Bring to the boil, then lower the heat and simmer gently for about 5 minutes, stirring occasionally, until the cream has reduced slightly.

4 Drain the pasta and return it to the pan. Add the grated Parmesan, then taste the sauce for seasoning and adjust if necessary. Pour it over the pasta. Toss quickly over a medium heat until the pasta is evenly coated with the sauce, then divide among four warmed bowls and serve immediately.

VARIATIONS

- Use spaghettini or spaghetti, or even small pasta shapes, such as fusilli, farfalle or orecchiette.
- For an even tangier taste, add a little grated lemon rind to the sauce when you add the butter and the cream to the pan in Step 2.

PENNE WITH ARTICHOKES

ARTICHOKES ARE A VERY POPULAR VEGETABLE IN ITALY, AND ARE OFTEN USED IN SAUCES FOR PASTA. THIS SAUCE IS RICHLY FLAVOURED WITH GARLIC, TOMATOES, WINE AND LEMON JUICE – THE PERFECT DISH TO SERVE AS A FIRST COURSE WHEN GLOBE ARTICHOKES ARE IN SEASON.

SERVES SIX

INGREDIENTS

- juice of 1 lemon
- 2 globe artichokes
- 30ml/2 tbsp olive oil
- 1 small fennel bulb, thinly sliced, with feathery tops reserved
- 1 onion, finely chopped
- 4 garlic cloves, finely chopped
- 1 handful fresh flat leaf parsley, coarsely chopped
- 400g/14oz can chopped plum tomatoes
- 150ml/¼ pint/⅔ cup dry white wine
- 350g/12oz/3 cups dried penne
- 10ml/2 tsp capers, chopped
- salt and ground black pepper
- freshly grated Parmesan cheese, to serve

1 Fill a large mixing bowl with cold water and add the juice of half a lemon. To prepare the artichokes, cut or break off the artichoke stalks, then pull off and discard the outer leaves until only the pale inner leaves that are almost white at the base remain.

2 Cut off the tops of these leaves so that the base remains. Cut the base in half lengthways, then prise the hairy choke out of the centre with the tip of the knife and discard. Cut the artichokes lengthways into 5mm/¼in slices, adding them immediately to the bowl of acidulated water to prevent them from discolouring.

3 Bring a large pan of salted water to the boil. Drain the artichokes and add them immediately to the water. Boil for 5 minutes, then drain and set aside.

4 Heat the oil in a large skillet or pan and add the fennel, onion, garlic and parsley. Cook over a low to medium heat, stirring frequently, for 10 minutes, or until the fennel has softened and is lightly coloured.

5 Add the tomatoes and wine, with salt and pepper to taste. Bring to the boil, stirring, then lower the heat, cover and simmer for 10–15 minutes. Stir in the artichokes, replace the lid and simmer for 10 minutes more. Meanwhile, add the pasta to a large pan of lightly salted boiling water and cook according to the instructions on the packet.

6 Drain the pasta, reserving a little of the cooking water. Add the capers to the sauce, stir well, then taste for seasoning. Add the remaining lemon juice.

7 Tip the pasta into a warmed, large serving bowl, pour the sauce over and toss thoroughly to mix, adding a little of the reserved cooking water if you like a thinner sauce. Serve immediately, garnished with the reserved fennel fronds. Hand around a bowl of grated Parmesan separately.

Main Courses

Citrus fruit is the perfect foil for rich meat and poultry dishes, such as the classic duck with orange, and the distinctive pairing of lamb and preserved lemon. It is also ideal for adding piquancy to lighter meats, such as chicken and veal. Freshly squeezed juice makes tangy marinades, and including fruit slices in oven-baked fish parcels is an easy way to add extra flavour. Fish is a popular choice for the health-conscious, and cooking it with citrus fruit is a fat-free way of ensuring its delicate flesh stays moist.

CHICKEN THIGHS WITH LEMON AND GARLIC

THIS RECIPE FEATURES CLASSIC FLAVOURINGS FOR CHICKEN – PUNGENT GARLIC, WHITE WINE AND FRESH LEMON. DIFFERENT VERSIONS OF THIS SUCCULENT DISH CAN BE FOUND IN BOTH SPAIN AND ITALY. THIS PARTICULAR RECIPE, HOWEVER, IS OF FRENCH ORIGIN.

SERVES FOUR

INGREDIENTS

- 600ml/1 pint/2½ cups chicken stock
- 20 large garlic cloves
- 25g/1oz/2 tbsp butter
- 15ml/1 tbsp olive oil
- 8 chicken thighs
- 1 lemon, peeled, pith removed and thinly sliced
- 30ml/2 tbsp plain (all-purpose) flour
- 150ml/¼ pint/⅔ cup dry white wine
- salt and ground black pepper
- chopped fresh parsley or basil, to garnish

1 Pour the stock into a pan and bring to the boil. Add the garlic, cover and simmer for 40 minutes.

2 Meanwhile, heat the butter and oil in a sauté or frying pan. Add the chicken thighs and cook over a low heat, turning occasionally until golden on all sides. With tongs or a slotted spoon, transfer them to an ovenproof dish. Preheat the oven to 190°C/375°F/Gas 5.

3 Strain the chicken stock and reserve it. Distribute the garlic and lemon slices among the chicken pieces. Add the flour to the fat in the pan in which the chicken was browned, and cook, stirring constantly, for 1 minute. Add the wine, stirring and scraping the base of the pan, then pour in the stock. Cook, still stirring, until the sauce has thickened and is smooth. Season to taste with a little salt and pepper.

4 Pour the sauce over the chicken, cover and transfer to the oven. Cook for 40–45 minutes until the chicken is tender and cooked through. If a thicker sauce is required, lift out the chicken pieces and reduce the sauce by boiling rapidly until it reaches the desired consistency. Sprinkle over the chopped parsley or basil and serve with boiled new potatoes or rice.

VARIATION

Substitute a bitter orange, such as Seville (Temple), for the lemon and dry vermouth for the white wine.

CHICKEN WITH PRESERVED LEMON AND OLIVES

THIS IS ONE OF THE MOST FAMOUS MOROCCAN DISHES. YOU MUST USE PRESERVED LEMON AS FRESH LEMON SIMPLY DOES NOT HAVE THE MELLOW FLAVOUR THAT THIS DISH REQUIRES. FOR A TRULY AUTHENTIC FLAVOUR, USE TAN-COLOURED MOROCCAN OLIVES.

SERVES FOUR

INGREDIENTS

- 30ml/2 tbsp olive oil
- 1 Spanish onion, chopped
- 3 garlic cloves
- 1cm/½in piece fresh root ginger, peeled and grated, or 2.5ml/½ tsp ground ginger
- 2.5–5ml/½–1 tsp ground cinnamon
- pinch of saffron threads
- 4 chicken quarters, preferably breast portions, halved if you like
- 750ml/1¼ pints/3 cups chicken stock
- 30ml/2 tbsp chopped fresh coriander (cilantro)
- 30ml/2 tbsp chopped fresh parsley
- 1 preserved lemon
- 115g/4oz/⅔ cup Moroccan tan olives
- salt and ground black pepper
- lemon slices and fresh coriander (cilantro) sprigs, to garnish

1 Heat the oil in a large flameproof casserole, add the onion and cook over a medium heat, stirring occasionally, for 6–8 minutes until lightly golden.

2 Meanwhile, crush the garlic and blend with the ginger, cinnamon, saffron and a little salt and pepper. Stir into the pan and cook for about 1 minute. Add the chicken and cook over a medium heat for 2–3 minutes, or until golden.

3 Add the stock and herbs, and bring to the boil. Cover and simmer for about 45 minutes until the chicken is tender.

4 Rinse the preserved lemon under cold running water, discard the flesh and cut the peel into small pieces. Stir into the pan with the olives, and simmer for a further 15 minutes, or until the chicken is very tender.

5 Transfer the chicken to a plate and keep warm. Bring the sauce to the boil and cook for 3–4 minutes, or until reduced and fairly thick. Pour it over the chicken and serve, garnished with lemon slices and coriander sprigs.

MEDITERRANEAN CHICKEN AND VEGETABLES

LEMON AND THYME ADD A FRAGRANCE TO CHICKEN ROASTED WITH PEPPER, AUBERGINE, FENNEL AND GARLIC IN THIS WONDERFUL FRENCH ALTERNATIVE TO A TRADITIONAL ROAST CHICKEN. FOR ADDITIONAL CITRUS APPEAL, USE LEMON THYME RATHER THAN THE ORDINARY VARIETY.

SERVES FOUR

INGREDIENTS

- 1.8–2kg/4–4½lb roasting chicken
- 150ml/¼ pint/⅔ cup extra virgin olive oil
- ½ lemon
- few sprigs of fresh thyme
- 450g/1lb small new potatoes
- 1 aubergine (eggplant), cut into 2.5cm/1in cubes
- 1 red (bell) pepper, seeded and quartered
- 1 fennel bulb, trimmed and quartered
- 8 large garlic cloves, unpeeled
- coarse salt and ground black pepper

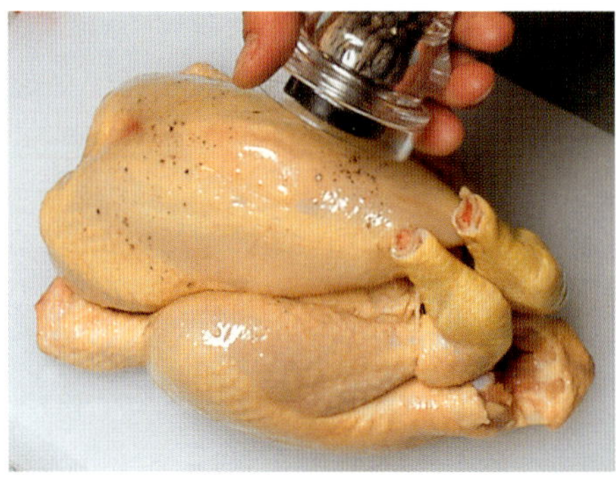

1 Preheat the oven to 200°C/400°F/ Gas 6. Rub the chicken with some of the oil and pepper. Place the lemon and two thyme sprigs inside it. Transfer to a roasting pan. Roast for 30 minutes.

2 Remove the chicken from the oven and season with salt. Turn it over and baste with the juices. Surround the bird with the potatoes, roll them in the pan juices, and return the pan to the oven.

3 After 30 minutes, add the aubergine, red pepper, fennel and garlic cloves to the pan. Drizzle with the remaining oil, and season with salt and pepper. Add the remaining thyme to the vegetables. Return the chicken to the oven, and cook for 30–50 minutes more, basting and turning the vegetables occasionally.

4 To find out if the chicken is cooked, push the tip of a sharp knife between the thigh and breast. If the juices run clear, it is done. Test the vegetables with a fork: they should be tender and just beginning to brown.

5 Serve the chicken and vegetables from the pan, or transfer the vegetables to a serving dish, joint the chicken and place it on top. Serve the skimmed juices in a gravy boat.

GRILLED POUSSINS WITH CITRUS GLAZE

THIS RECIPE IS SUITABLE FOR MANY KINDS OF SMALL BIRDS, INCLUDING PIGEONS, SNIPE AND PARTRIDGES. IT WOULD ALSO WORK WITH QUAIL, BUT DECREASE THE COOKING TIME AND SPREAD THE CITRUS MIXTURE OVER, RATHER THAN UNDER, THE FRAGILE SKIN.

SERVES FOUR

INGREDIENTS

- 2 poussins, about 675g/1½lb each
- 50g/2oz/¼ cup unsalted (sweet) butter, softened
- 30ml/2 tbsp olive oil
- 2 garlic cloves, crushed
- 2.5ml/½ tsp dried thyme
- 1.5ml/¼ tsp cayenne pepper, or to taste
- grated rind and juice of 1 lemon
- grated rind and juice of 1 lime
- 30ml/2 tbsp clear honey
- salt and ground black pepper
- fresh dill sprigs, to garnish
- tomato salad, to serve

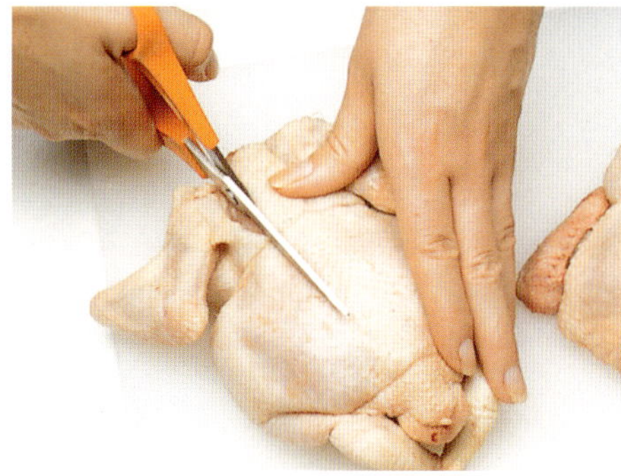

1 Using kitchen scissors, cut along both sides of the backbone of each bird; remove and discard. Cut the birds in half along the breast bone. Using a rolling pin, press down to flatten.

2 Beat the butter in a small bowl with a wooden spoon, then beat in 15ml/1 tbsp of the olive oil, the garlic, thyme, cayenne, salt and pepper, half the lemon and lime rind and 15ml/1 tbsp each of the lemon and lime juice.

3 Using your fingertips, carefully loosen the skin of each poussin breast. With a round-bladed knife or small spatula, spread the butter mixture between the skin and breast meat.

4 Preheat the grill (broiler) and line a grill pan with foil. In a small bowl, mix together the remaining olive oil, lemon and lime juices and the honey. Place the bird halves, skin-side up, on the grill pan and brush with the juice and honey mixture.

5 Grill (broil) for 10–12 minutes, basting twice with the juices. Turn over and grill for 7–10 minutes, basting once, until the juices run clear when the thigh is pierced. Serve with tomato salad, garnished with dill.

Duck with Orange Sauce

This traditional sauce of fresh oranges and orange liqueur is always a favourite. As commercially raised ducks tend to be much fatter than wild ducks, initial slow cooking and pricking the duck's skin helps to draw out excess fat.

SERVES THREE TO FOUR

INGREDIENTS

- 2kg/4½lb duck
- 2 oranges
- 90g/3½oz/½ cup caster (superfine) sugar
- 90ml/6 tbsp white wine vinegar or cider vinegar
- 120ml/4fl oz/½ cup Grand Marnier or other orange liqueur
- salt and ground black pepper
- watercress and orange slices, to garnish

1 Preheat the oven to 150°C/300°F/Gas 2. Prick the duck's skin all over with a fork. Season the duck inside and out with salt and pepper and tie the legs with string.

2 Place the duck on a rack in a large roasting pan. Cover with foil and roast for 1½ hours. With a vegetable peeler, remove the rind in wide strips from the oranges, then stack two or three strips at a time and slice into very thin julienne strips. Squeeze the juice from the oranges.

3 Place the sugar and vinegar in a small, heavy pan and stir to dissolve the sugar. Boil over a high heat, without stirring, until the mixture is a rich caramel colour.

4 Remove the pan from the heat and add the orange juice, pouring it down the side of the pan. Swirl to blend, bring back to the boil and add the orange rind and liqueur. Simmer for 3 minutes.

5 Remove the duck from the oven and carefully pour off all the fat from the pan. Raise the oven temperature to 200°C/400°F/Gas 6.

6 Roast the duck, uncovered, for about 30 minutes, basting three or four times with the caramel and orange mixture, until the duck is golden brown and the juices run clear when the thigh is pierced with a knife.

7 Pour the juices from the cavity into the roasting pan and transfer the duck to a carving board. Cover loosely with foil and leave to stand for 15 minutes. Skim the fat from the roasting juices, and pour the juices into the pan with the rest of the caramel mixture and simmer. Serve the duck, with the sauce, garnished with watercress and orange slices.

COOK'S TIP

For crisp skin, the duck must be dry before roasting. Leave, uncovered, in a cool place for 2 hours before roasting.

PHEASANT WITH LEMON AND SAGE

A POPULAR GAME BIRD, YOUNG PHEASANTS ARE IDEALLY SUITED TO GRILLING AND ROASTING. CITRUS RIND AND SAGE IS TUCKED BENEATH THE SKIN SO THAT THEIR FLAVOURS PERMEATE THE MEAT WHILE IT COOKS. WITH THE MUSTARD AND BRANDY SAUCE ACCOMPANIMENT, THIS IS A TASTY MEAL INDEED.

SERVES FOUR

INGREDIENTS

- 2 pheasants, about 450g/1lb each
- 1 lemon
- 60ml/4 tbsp chopped fresh sage leaves
- 3 shallots
- 5ml/1 tsp Dijon mustard
- 15ml/1 tbsp brandy or dry sherry
- 150ml/¼ pint/⅔ cup crème fraîche
- salt and ground black pepper
- lemon wedges and sage sprigs, to garnish

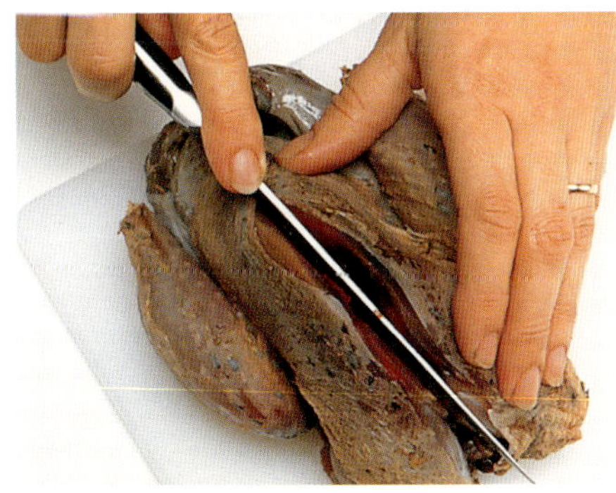

1 Place the pheasants, breast-side up, on a chopping board and cut them in half lengthways, using poultry shears or a sharp knife.

2 Grate the rind from half the lemon and thinly slice the other half. Place the lemon rind and half the chopped sage in a small bowl and mix well. Preheat the grill (broiler) or light the charcoal for a barbecue.

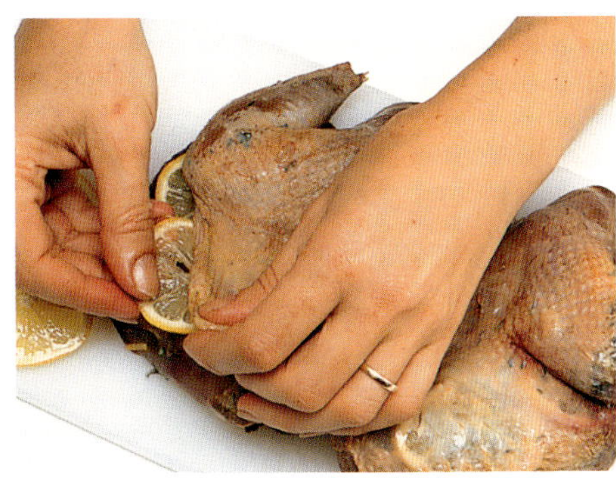

3 Loosen the skin on the breast and legs of the pheasants and push a little of the sage mixture under each. Tuck the lemon slices under the skin, then smooth the skin back firmly.

4 Place each half-pheasant under the grill or on a medium-hot barbecue rack and cook for 25–30 minutes, turning them once.

5 Meanwhile, place the shallots under the grill or on the barbecue and cook for 10–12 minutes, turning occasionally, until the skin is blackened. Peel off the skins, chop the flesh and mash it with the mustard and brandy or sherry. Stir in the crème fraîche, add the remaining sage and season to taste. Garnish the pheasants with lemon wedges and sage sprigs and serve with the sauce.

Veal Escalopes with Lemon

Popular in Italian restaurants, this dish is very easy to make at home. White vermouth and lemon juice make the perfect sauce for the delicately flavoured meat.

SERVES FOUR

INGREDIENTS

- 4 veal escalopes (US veal scallops)
- 30–45ml/2–3 tbsp plain (all-purpose) flour
- 50g/2oz/¼ cup butter
- 60ml/4 tbsp olive oil
- 60ml/4 tbsp Italian dry white vermouth or dry white wine
- 45ml/3 tbsp lemon juice
- salt and ground black pepper
- lemon wedges, grated lemon rind and fresh parsley, to garnish
- salad, to serve

VARIATIONS

- Use skinless, boneless chicken breast portions instead of the veal. If they are thick, cut them in half before pounding.
- Substitute thin slices of pork fillet (tenderloin) for the veal and use orange juice instead of lemon.

1 Put each veal escalope between two sheets of clear film (plastic wrap) and pound with the side of a rolling pin or the smooth side of a meat mallet until very thin.

2 Cut the pounded escalopes in half or quarters. Season the flour with a little salt and pepper and use it to coat the escalopes on both sides. Shake off any excess.

3 Melt the butter with half the oil in a large, heavy frying pan until sizzling. Add as many escalopes as the pan will hold. Cook over a medium to high heat for about 2 minutes on each side until lightly coloured. Remove with a spatula and keep hot. Add the remaining oil to the pan and cook the remaining veal escalopes in the same way.

4 Remove the pan from the heat and add the vermouth or wine and the lemon juice. Stir vigorously to mix well with the pan juices, then return the pan to the heat. Return all the veal escalopes to the pan.

5 Spoon the sauce over the veal. Shake the pan over a medium heat until the escalopes are coated in the sauce and heated through. Serve with a fresh salad, garnished with lemon wedges, lemon rind and sprinkled with parsley.

OSSO BUCO WITH GREMOLATA

SLOW-COOKED VEAL IN A RICH SAUCE IS SERVED WITH A ZESTY GREMOLATA OF PARSLEY, LEMON AND GARLIC. THIS RICH AND HEARTY DISH FROM MILAN IS EXCELLENT SERVED WITH BOILED RICE.

SERVES FOUR

INGREDIENTS

- 30ml/2 tbsp plain (all-purpose) flour
- 4 pieces of osso buco (veal shanks)
- 2 small onions
- 30ml/2 tbsp olive oil
- 1 large celery stick, finely chopped
- 1 carrot, finely chopped
- 2 garlic cloves, finely chopped
- 400g/14oz can chopped tomatoes
- 300ml/½ pint/1¼ cups dry white wine
- 300ml/½ pint/1¼ cups chicken or veal stock
- 1 strip of thinly pared lemon rind
- 2 bay leaves, plus extra to garnish
- salt and ground black pepper

For the gremolata

- 30ml/2 tbsp finely chopped fresh flat leaf parsley
- finely grated rind of 1 lemon
- 1 garlic clove, finely chopped

1 Preheat the oven to 160°C/325°F/Gas 3. Season the flour with salt and pepper and spread it out in a shallow dish. Add the pieces of osso buco and turn them in the flour until evenly coated. Shake off any excess flour.

2 Slice one of the onions and separate it into rings. Heat the olive oil in a large flameproof casserole, then add the veal pieces and the onion rings. Cook over a medium heat until the veal is browned on both sides. Remove the pieces of veal with tongs and set aside on kitchen paper to drain.

3 Chop the remaining onion and add it to the casserole with the celery, carrot and garlic. Stir, scraping the base of the casserole to incorporate the cooking juices and sediment. Cook over a low heat, stirring frequently, for about 5 minutes, or until the vegetables begin to soften slightly.

4 Add the chopped tomatoes, wine, stock, lemon rind and bay leaves, then season to taste with salt and pepper. Bring to the boil, stirring constantly. Return the veal to the casserole and coat with the sauce. Cover and cook in the oven for 2 hours, or until the veal feels tender when pierced with a fork.

5 To make the gremolata, mix together the chopped parsley, lemon rind and garlic in a small bowl.

6 Remove the casserole from the oven and lift out and discard the strip of lemon rind and the bay leaves. Taste the sauce and adjust the seasoning, if necessary. Sprinkle with the gremolata, garnish with extra bay leaves and serve the osso buco immediately.

Loin of Pork with Cashew Nut and Orange Stuffing

THE ORANGES AND CASHEW NUTS ADD CONTRASTING FLAVOURS AND TEXTURES TO THIS DELICIOUS STUFFING, AND COMBINE WELL WITH THE BROWN RICE.

SERVES SIX

INGREDIENTS

- 1.3–1.6kg/3–3½lb boned pork loin
- 15ml/1 tbsp plain (all-purpose) flour
- 300ml/½ pint/1¼ cups dry white wine
- salt and ground black pepper
- fresh rosemary sprigs and orange slices, to garnish

For the stuffing

- 25g/1oz/2 tbsp butter
- 1 small onion, finely chopped
- 75g/3oz/scant ½ cup brown basmati rice, soaked for 30 minutes
- 350ml/12fl oz/scant 1½ cups chicken stock
- 50g/2oz/⅓ cup cashew nuts
- 1 orange
- 50g/2oz/⅓ cup sultanas (golden raisins)

1 First prepare the stuffing. Melt the butter in a frying pan and cook the onion for 2–3 minutes until softened. Drain the rice, add it to the pan and cook for 1 minute, then pour in the stock and bring to the boil. Stir, then lower the heat, cover and simmer for 35 minutes, or until the rice is tender and the liquid has been absorbed. Preheat the oven to 220°C/425°F/Gas 7.

2 Meanwhile, open out the pork loin and cut two lengthways slits through the meat, but without cutting right through. Turn the meat over. Remove any excess fat, but leave a good layer.

3 Spread out the cashew nuts for the stuffing in a roasting pan and roast for 2–4 minutes, or until golden. Set aside to cool, then chop coarsely in a food processor or blender. Leave the oven switched on.

4 Grate 5ml/1 tsp of the orange rind into a bowl. Using a sharp knife, cut a thin slice of peel and pith from each end of the orange. Place the orange cut side down on a plate and cut off the peel and pith in strips. Remove any remaining pith. Cut out each segment leaving the membrane behind. Squeeze the remaining juice from the membrane. Chop the segments coarsely.

5 Add the chopped orange segments to the cooked rice and stir in the orange rind, roast cashew nuts and sultanas. Season well with salt and pepper, then stir in 15–30ml/1–2 tbsp of the reserved orange juice. Do not worry if the rice does not bind – it should have a fairly loose consistency.

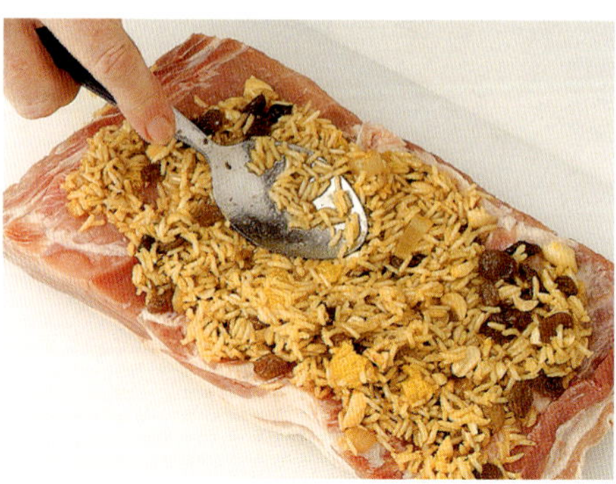

6 Spread a generous layer of stuffing along the centre of the pork. Put any leftover stuffing in an ovenproof bowl.

7 Roll up the loin and firmly tie with kitchen string. Rub salt and pepper into the meat and place it in a roasting pan. Cook for 15 minutes, then lower the oven temperature to 180°C/350°F/Gas 4. Roast for 2–2¼ hours more, or until the juices run clear. Heat any extra stuffing in the covered bowl alongside the meat for the final 15 minutes.

8 Transfer the meat to a serving plate and keep warm. Stir the flour into the juices remaining in the roasting pan, cook for 1 minute, then stir in the white wine. Bring to the boil, stirring until thickened, then strain into a gravy boat.

9 Remove the string from the meat before carving. Stud the pork with the rosemary and garnish with the orange slices. Serve with the gravy and any extra stuffing.

SPICY VENISON CASSEROLE WITH ORANGES AND CRANBERRIES

LOW IN FAT BUT HIGH IN FLAVOUR, VENISON IS AN EXCELLENT CHOICE FOR HEALTHY, YET RICH, CASSEROLES. CRANBERRIES AND ORANGE BRING A FESTIVE FRUITINESS TO THIS SPICY RECIPE. IT IS DELICIOUS SERVED WITH SMALL BAKED POTATOES AND GREEN VEGETABLES.

SERVES FOUR

INGREDIENTS

- 30ml/2 tbsp olive oil
- 1 onion, chopped
- 2 celery sticks, sliced
- 10ml/2 tsp ground allspice
- 15ml/1 tbsp plain (all-purpose) flour
- 675g/1½ lb stewing venison, cubed
- 225g/8oz fresh or frozen cranberries
- grated rind and juice of 1 orange
- 900ml/1½ pints/3¾ cups beef or venison stock
- salt and ground black pepper

1 Heat the olive oil in a flameproof casserole. Add the onion and celery and cook for 5 minutes until softened.

2 Meanwhile, mix the ground allspice with the flour and either spread the mixture out on a large plate or place in a large plastic bag. Toss a few pieces of venison at a time (to prevent them from becoming soggy) in the flour mixture until they are all lightly coated.

3 When the onion and celery are softened, remove from the casserole using a slotted spoon and set aside. Add the venison pieces to the casserole in batches and cook until browned and sealed on all sides.

4 Add the cranberries, orange rind and juice to the casserole, pour in the beef or venison stock, and stir well. Return the vegetables and all the venison to the casserole and heat until simmering, then cover tightly and reduce the heat. Simmer, stirring occasionally, for about 45 minutes, or until the meat is tender.

5 Season the venison casserole to taste with salt and pepper before serving.

VARIATIONS

- Farmed venison is increasingly easy to find and is available from good butchers and many supermarkets. It makes a rich stew, but lean pork or braising steak could be used in place of the venison.
- You could replace the cranberries with pitted and halved ready-to-eat prunes and, for an extra rich flavour, use stout instead of half the meat stock.

LAMB TAGINE WITH ARTICHOKES AND PRESERVED LEMON

A TAGINE IS A SLOWLY SIMMERED STEW TRADITIONALLY COOKED IN AN EARTHENWARE POT CALLED A TAGINE. IN THIS RECIPE, THE MEAT IS MARINATED IN A BLEND OF FLAVOURS BEFORE COOKING, THEN PRESERVED LEMON AND ARTICHOKES ARE ADDED TO PROVIDE THEIR DISTINCTIVE FLAVOURS.

SERVES FOUR TO SIX

INGREDIENTS

- 675g/1½lb leg of lamb, trimmed and cut into cubes
- 2 onions, very finely chopped
- 2 garlic cloves, crushed
- 60ml/4 tbsp chopped fresh parsley
- 60ml/4 tbsp chopped fresh coriander (cilantro)
- good pinch of ground ginger
- 5ml/1 tsp ground cumin
- 90ml/6 tbsp olive oil
- 350–400ml/12–14fl oz/1½–1⅔ cups water or stock
- 1 preserved lemon
- 400g/14oz can artichoke hearts, drained and halved
- 15ml/1 tbsp chopped fresh mint, plus extra sprigs to garnish
- 1 egg, beaten (optional)
- salt and ground black pepper
- couscous, to serve

1 Place the meat in a shallow dish. Mix together the onions, garlic, parsley, coriander, ginger, cumin, seasoning and olive oil. Stir into the meat, cover with clear film (plastic wrap) and set aside to marinate for 3–4 hours or overnight.

2 Heat a large, heavy pan and add the meat and marinade. Cook over a high heat for 5–6 minutes, then stir in enough water or stock to just cover it. Bring to the boil, cover and simmer for 45–60 minutes, or until the meat is just tender.

3 Rinse the preserved lemon under cold water, discard the flesh and cut the peel into pieces. Stir into the meat and simmer for a further 15 minutes, then add the artichoke hearts and mint.

4 Simmer for a few minutes. If you wish to thicken the sauce, remove the pan from the heat and stir in some or all of the egg. Garnish with mint and serve with couscous.

Scottish Salmon with Herb Butter

Cooking salmon in a parcel seals in all those lovely flavours and juices. It is ideal when using a subtle dill and lemon butter that imparts a delicate fragrance to the fish.

SERVES FOUR

INGREDIENTS

- 50g/2oz/¼ cup butter, softened
- finely grated rind of ½ small lemon
- 15ml/1 tbsp lemon juice
- 15ml/1 tbsp chopped fresh dill
- 4 salmon steaks
- 2 lemon slices, halved
- 4 fresh dill sprigs
- salt and ground black pepper
- new potatoes and sugar snap peas, to serve

COOK'S TIP

Other fresh herbs could be used to flavour the butter – try mint, fennel fronds, lemon balm, parsley or oregano instead of the dill.

1 Place the butter, grated lemon rind, lemon juice and chopped dill in a small bowl, then season to taste with salt and plenty of black pepper. Mix together with a fork until thoroughly blended and smooth.

2 Spoon the butter on to a small piece of baking parchment and then roll up, smoothing with your hands into a sausage shape. Twist the ends tightly, wrap in clear film (plastic wrap) and put in the freezer for 20 minutes until firm.

3 Meanwhile, preheat the oven to 190°C/375°F/Gas 5. Cut out four squares of foil large enough to encase the salmon steaks, and grease lightly. Place a salmon steak in the centre of each one.

4 Remove the butter from the freezer and slice into eight rounds. Place two rounds on top of each salmon steak with a halved lemon slice in the centre and a sprig of dill on top. Lift up the edges of the foil and firmly crinkle them together until well sealed.

5 Lift the parcels on to a baking sheet and bake for about 20 minutes. When cooked, place the unopened parcels on warmed plates. Open the parcels and slide the fish and its juices on to the plates. Serve immediately.

PAN-FRIED RED MULLET WITH CITRUS

RED MULLET IS POPULAR ALL OVER THE MEDITERRANEAN. THIS ITALIAN RECIPE COMBINES IT WITH ORANGES AND LEMONS, WHICH GROW IN ABUNDANCE THERE.

SERVES FOUR

INGREDIENTS

- 2 oranges, such as Navelina
- 4 red mullet or snapper, about 225g/8oz each, filleted
- 90ml/6 tbsp olive oil
- 10 black peppercorns, crushed
- 1 lemon
- 30ml/2 tbsp plain (all-purpose) flour
- 15g/½oz/1 tbsp butter
- 2 drained canned anchovies, chopped
- 60ml/4 tbsp shredded fresh basil
- salt and ground black pepper

1 Using a sharp knife, cut the peel and pith from one of the oranges, then slice the flesh into rounds. Squeeze the juice from the other orange into a small bowl, cover and set aside.

2 Place the fish fillets in a shallow dish in a single layer. Pour over the olive oil and sprinkle with the crushed black peppercorns. Lay the orange slices on top of the fish. Cover the dish and leave to marinate in the refrigerator for at least 4 hours.

3 Halve the lemon. Cut the skin and pith from one half using a small sharp knife and slice thinly and reserve. Squeeze the juice from the other half.

VARIATION
If you prefer, use other fish fillets for this dish, such as lemon sole, monkfish, haddock or hake.

4 Carefully lift the fish out of the marinade, and pat dry with kitchen paper. Reserve the marinade and orange slices. Season the fish on both sides with salt and pepper, then dust lightly with flour.

5 Heat 45ml/3 tbsp of the marinade in a frying pan. Add the fish and cook for 2 minutes on each side. Remove from the pan and keep warm. Discard the marinade that is left in the pan.

6 Melt the butter in the pan with any of the remaining original marinade. Add the anchovies and cook over a low heat until completely softened.

7 Stir in the orange and lemon juice, then check and adjust the seasoning if necessary. Simmer until the sauce has slightly reduced. Stir in the basil. Pour the sauce over the fish, garnish with the reserved orange slices and the lemon slices, and serve immediately.

SALMON FISHCAKES

THE SECRET OF A GOOD FISHCAKE IS TO MAKE IT WITH FRESHLY PREPARED FISH AND POTATOES, HOME-MADE BREADCRUMBS AND PLENTY OF INTERESTING FLAVOURINGS – WHOLEGRAIN MUSTARD, FRESH HERBS AND LEMON ARE THE PERFECT COMBINATION.

SERVES FOUR

INGREDIENTS

- 450g/1lb cooked salmon fillet
- 450g/1lb freshly cooked potatoes, mashed
- 25g/1oz/2 tbsp butter, melted
- 10ml/2 tsp wholegrain mustard
- 15ml/1 tbsp each chopped fresh dill and chopped fresh parsley
- grated rind and juice of ½ lemon
- 15g/½oz/2 tbsp plain (all-purpose) flour
- 1 egg, lightly beaten
- 150g/5oz/2 cups dried breadcrumbs
- 60ml/4 tbsp sunflower oil
- salt and ground black pepper
- rocket (arugula) leaves, chives and lemon wedges, to garnish

1 Flake the cooked salmon, discarding any skin and bones. Put the flesh in a bowl with the mashed potato, melted butter and wholegrain mustard, and mix well. Stir in the dill and parsley, lemon rind and juice. Season to taste with salt and pepper.

2 Divide the mixture into eight portions and shape each into a ball, then flatten into a thick disc. Coat the fishcakes first in flour, then in egg and finally in breadcrumbs, making sure that they are evenly coated.

3 Heat the oil in a frying pan until it is very hot. Fry the fishcakes in batches until golden brown and crisp all over. As each batch is ready, drain on kitchen paper and keep hot. Garnish with rocket leaves, chives and lemon wedges and serve immediately.

VARIATION

Almost any fresh white or hot-smoked fish is suitable; smoked cod and haddock are particularly good.

PAN-FRIED SOLE WITH LEMON

THE DELICATE FLAVOUR AND TEXTURE OF SOLE IS BROUGHT OUT IN THIS SIMPLE, CLASSIC RECIPE USING TART CAPERS AND LEMON JUICE. LEMON SOLE IS USED HERE BECAUSE IT IS EASIER TO OBTAIN – AND LESS EXPENSIVE – THAN DOVER SOLE.

SERVES TWO

INGREDIENTS
- 30–45ml/2–3 tbsp plain (all-purpose) flour
- 4 lemon sole fillets
- 45ml/3 tbsp olive oil
- 50g/2oz/¼ cup butter
- 60ml/4 tbsp lemon juice
- 30ml/2 tbsp rinsed bottled capers
- salt and ground black pepper
- fresh flat leaf parsley and lemon wedges, to garnish

COOK'S TIP

It is important to cook the pan juices to the right colour after removing the fish. Too pale, and they will taste insipid, too dark, and they may taste bitter. Take great care not to be distracted at this point so that you can watch the colour of the juices change to a golden brown.

1 Season the flour with salt and pepper. Coat the fish evenly on both sides. Put the oil and half the butter in a large pan and heat until foaming. Add two sole fillets and fry over a medium heat for 2–3 minutes on each side.

2 Lift out the sole fillets with a spatula and place on a warmed serving platter. Keep hot. Fry the remaining sole fillets.

3 Remove the pan from the heat and add the lemon juice and remaining butter. Return to a high heat and stir until the juices are sizzling and are turning golden brown. Remove from the heat and stir in the capers.

4 Pour the pan juices over the sole and season. Garnish with the parsley and lemon wedges and serve immediately.

MONKFISH WITH CITRUS MARINADE

A MEATY FISH THAT KEEPS ITS SHAPE WELL, MONKFISH IS IDEAL FOR COOKING ON A BARBECUE OR GRILL. A TRIO OF CITRUS FRUITS IMPART THEIR DELICATE FLAVOURS TO PEPPERED MONKFISH FILLETS.

SERVES FOUR

INGREDIENTS

- 2 monkfish tails, about 350g/12oz each
- 1 lime
- 1 lemon
- 2 Valencia or blood oranges
- handful of fresh thyme sprigs
- 30ml/2 tbsp olive oil
- 15ml/1 tbsp mixed peppercorns, coarsely crushed
- salt and ground black pepper

VARIATION

You can also use this marinade for monkfish kebabs.

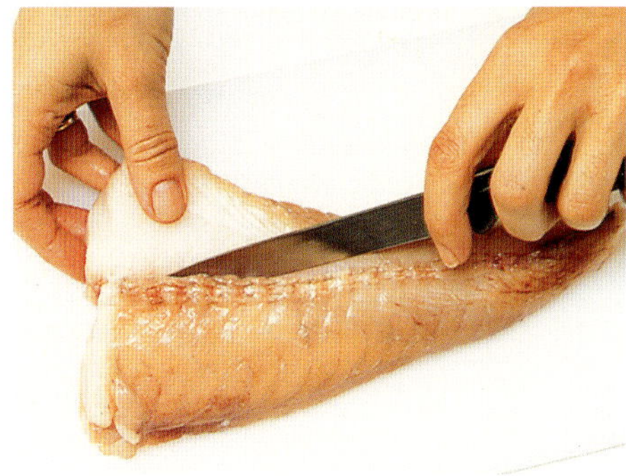

1 Remove any skin or membrane from the monkfish tails. Cut carefully down one side of the backbone, sliding the knife between the bone and flesh, and remove the fillet on one side.

2 Turn and repeat on the other side, to remove the second fillet. Repeat on the second tail to produce four fillets in all. Lay the fillets out flat.

3 Cut two slices from each fruit and arrange them over two of the fillets. Add a few sprigs of thyme and season well. Finely grate the rind from the remaining fruit and sprinkle it over the fish.

4 Lay the other two fillets on top and tie them firmly at intervals, with fine cotton string. Place in a wide dish.

5 Squeeze the juice from the fruit, mix it with the oil, and season. Spoon over the fish. Cover and leave to marinate for 1 hour, turning and spooning the marinade over it occasionally.

6 Drain the fish, pat dry with kitchen paper and sprinkle with the crushed peppercorns. Reserve the marinade. Cook the fish on a medium-hot barbecue, or under a grill (broiler), for about 20 minutes, basting with the marinade and turning occasionally.

GRILLED HALIBUT WITH SAUCE VIERGE

TOMATOES, CAPERS, ANCHOVIES, HERBS AND LEMON MAKE A VIBRANT SAUCE THAT IS PERFECT FOR HALIBUT, BUT IT IS ALSO SO VERSATILE IT WILL SUIT ANY THICK WHITE FISH FILLETS.

SERVES FOUR

INGREDIENTS

- 105ml/7 tbsp olive oil
- 2.5ml/½ tsp fennel seeds
- 2.5ml/½ tsp celery seeds
- 5ml/1 tsp mixed peppercorns
- 5ml/1 tsp fresh thyme leaves, chopped
- 5ml/1 tsp fresh rosemary leaves, chopped
- 5ml/1 tsp fresh oregano or marjoram leaves, chopped
- 675–800g/1½–1¾lb middle cut of halibut, about 3cm/1¼in thick, cut into four pieces
- coarse sea salt
- shredded lettuce and lemon wedges, to serve

For the sauce

- 2 tomatoes
- 105ml/7 tbsp extra virgin olive oil
- juice of 1 lemon
- 1 garlic clove, finely chopped
- 5ml/1 tsp small capers
- 2 drained canned anchovy fillets, chopped
- 5ml/1 tsp chopped fresh chives
- 15ml/1 tbsp shredded fresh basil leaves
- 15ml/1 tbsp chopped fresh chervil

1 Plunge the tomatoes into boiling water for 30 seconds, then refresh in cold water. Peel off the skins, remove the seeds and dice the flesh. Set aside.

2 Heat a ridged griddle or preheat the grill (broiler) to high. Brush the griddle or grill pan with a little of the olive oil.

3 Meanwhile, mix the fennel and celery seeds with the peppercorns in a mortar. Crush with a pestle, and then stir in sea salt to taste. Spoon the mixture into a large, flat dish and stir in the herbs and the remaining olive oil.

VARIATION

Try turbot, brill and John Dory, or even humbler fish, like cod or haddock, with this sauce.

4 Add the halibut pieces to the olive oil and herb mixture, turning them to coat thoroughly, then arrange them with the dark skin uppermost on the oiled griddle or grill pan. Cook or grill (broil) for about 7 minutes, turning once, or until the fish is cooked through and the skin has browned.

5 Combine all the sauce ingredients, except the fresh herbs, in a pan and heat gently until warm but not hot. Stir in the chives, basil and chervil.

6 Place the halibut on four plates and spoon the sauce over the fish. Serve with the lettuce and lemon wedges.

Hot Desserts

When people think about hot fruit desserts, citrus fruits do not immediately spring to mind, so the recipes featured here will probably come as a delightful surprise. There are, of course, traditional favourites – crêpes flambéed in orange liqueur are still a popular offering in Parisian-style bistros, while children never cease to enjoy the apparently magical surprise of a light sponge pudding swimming in a lovely, lemony syrup. However, there are lots of unexpected treats in store, from citrus-flavoured pies and tarts to familiar rice puddings given a tangy makeover.

Apricots with Citrus Almond Paste

Take advantage of the short apricot season by making this charming apricot and almond dessert, delicately scented with lemon juice and orange flower water.

SERVES SIX

INGREDIENTS

- 75g/3oz/6 tbsp caster (superfine) sugar
- 30ml/2 tbsp lemon juice
- 300ml/½ pint/1¼ cups water
- 115g/4oz/1 cup ground almonds
- 50g/2oz/½ cup icing (confectioners') sugar
- a little orange flower water
- 25g/1oz/2 tbsp unsalted (sweet) butter, melted
- 2.5ml/½ tsp almond essence (extract)
- 900g/2lb fresh apricots
- fresh mint sprigs, to decorate

1 Preheat the oven to 180°C/350°F/ Gas 4. Place the sugar, lemon juice and water in a small pan and bring to the boil, stirring occasionally until the sugar has all dissolved. Simmer gently for 5–10 minutes to make a thin syrup.

2 Place the ground almonds, icing sugar, orange flower water, butter and almond essence in a bowl and blend together to make a smooth paste.

3 Wash the apricots and then make a slit in the flesh and ease out the stone (pit). Take small pieces of the almond paste, roll into balls and press one into each of the apricots.

4 Arrange the stuffed apricots in a shallow ovenproof dish and carefully pour the sugar syrup around them. Cover with foil and bake in the oven for 25–30 minutes.

5 Serve the apricots with a little of the syrup, and decorated with sprigs of fresh mint.

COOK'S TIP

Always use a heavy pan when making syrup and stir constantly with a wooden spoon until the sugar has completely dissolved. Do not let liquid come to the boil before it has dissolved, or the result will be grainy.

FIGS AND PEARS IN HONEY

A STUNNINGLY SIMPLE DESSERT USING FRESH FIGS AND PEARS SCENTED WITH THE WARM FRAGRANCES OF CINNAMON AND CARDAMOM AND DRENCHED IN A LEMON AND HONEY SYRUP.

SERVES FOUR

INGREDIENTS

- 1 lemon
- 90ml/6 tbsp clear honey
- 1 cinnamon stick
- 1 cardamom pod
- 2 pears
- 8 fresh figs, halved

COOK'S TIPS

- Leave the peel on the pears or discard, depending on your preference.
- Figs vary in colour from pale green and yellow to dark purple. When buying, look for firm fruit without any bruises or blemishes. A ripe fig will yield gently in your hand without pressing.
- It is best to use pale green or light beige cardamom pods, rather than the coarser dark brown ones.

1 Pare the rind from the lemon using a zester. Alternatively, use a vegetable peeler and then cut into very thin strips.

2 Place the lemon rind, honey, cinnamon stick, cardamom pod and 350ml/12fl oz/1½ cups water in a heavy pan and boil, uncovered, for about 10 minutes until reduced by about half.

3 Cut the pears into eighths, discarding the cores. Place in the syrup, add the figs and simmer for about 5 minutes, or until the fruit is tender.

4 Transfer the fruit to a serving bowl. Continue cooking the liquid until syrupy, then discard the cinnamon stick and pour over the figs and pears. Serve.

Surprise Lemon Pudding

Although all the ingredients are mixed together, during cooking a tangy lemon sauce forms beneath a light topping, making this lemon pudding a tasty surprise.

Serves Four

Ingredients
- 75g/3oz/6 tbsp butter
- 175g/6oz/¾ cup soft light brown sugar
- 4 eggs, separated
- grated rind and juice of 4 lemons
- 50g/2oz/½ cup self-raising (self-rising) flour
- 120ml/4fl oz/½ cup milk

VARIATION

This pudding is also delicious made with oranges instead of lemons.

1 Preheat the oven to 180°C/350°F/ Gas 4. Butter an 18cm/7in soufflé dish and stand it in a roasting pan.

2 Beat the butter and sugar together in a large bowl until pale and very fluffy. Beat in one egg yolk at a time, beating well after each addition and gradually beating in the lemon rind and juice until well mixed; do not worry if the mixture curdles a little.

3 Sift the flour and stir it into the lemon mixture until well mixed, then gradually stir in the milk.

4 Whisk the egg whites in a separate bowl until stiff, but not dry, then lightly, but thoroughly, fold into the lemon mixture in three batches. Carefully pour the mixture into the soufflé dish, then pour boiling water into the roasting pan.

5 Bake the pudding in the middle of the oven for 45 minutes, or until golden on top. Dust with icing (confectioners') sugar and serve immediately.

COOK'S TIP

When whisking egg whites, use a grease-free bowl and make sure that there are no traces of yolk.

ORANGES IN HOT COFFEE SYRUP

THIS RECIPE MAKES A MOUTHWATERING DESSERT AND ALSO WORKS WELL WITH MOST CITRUS FRUITS; TRY SWEET CLEMENTINES AS AN ALTERNATIVE TO THE ORANGES.

SERVES SIX

INGREDIENTS

- 6 medium oranges
- 200g/7oz/1 cup sugar
- 50ml/2fl oz/¼ cup cold water
- 100ml/3½fl oz/scant ½ cup boiling water
- 100ml/3½fl oz/scant ½ cup fresh strong brewed coffee
- 50g/2oz/⅓ cup pistachio nuts, chopped (optional)

COOK'S TIP

Choose a pan in which the oranges will just fit in a single layer.

1 Finely pare the rind from one orange, shred and reserve the rind. Peel the remaining oranges. Cut each orange crossways into slices, then re-form and hold in place with a cocktail stick (toothpick) through the centre.

2 Put the sugar and cold water in a pan. Heat gently, stirring constantly, until the sugar dissolves, then bring to the boil and cook until the syrup turns pale gold.

3 Remove from the heat and carefully pour the boiling water into the pan. Return to the heat until the syrup has dissolved in the water. Stir in the coffee.

4 Add the oranges and the shredded rind to the coffee syrup. Simmer for 15–20 minutes, turning the oranges once during cooking. Sprinkle with pistachio nuts, if using, and serve hot.

Crêpes with Orange Sauce

This is one of the best-known French desserts and is easy to do at home. You can make the crêpes in advance, and then coat them in the tangy orange sauce at the last minute.

SERVES SIX

INGREDIENTS

- 115g/4oz/1 cup plain (all-purpose) flour
- 1.5ml/¼ tsp salt
- 25g/1oz/2 tbsp caster (superfine) sugar
- 2 eggs, lightly beaten
- about 250ml/8fl oz/1 cup milk
- about 60ml/4 tbsp water
- 30ml/2 tbsp orange flower water, Cointreau or orange liqueur
- 25g/1oz/2 tbsp unsalted (sweet) butter, melted, plus extra for frying

For the sauce

- 75g/3oz/6 tbsp unsalted (sweet) butter
- 50g/2oz/¼ cup caster (superfine) sugar
- grated rind and juice of 1 large orange, such as Jaffa
- grated rind and juice of 1 lemon
- 150ml/¼ pint/⅔ cup freshly squeezed orange juice
- 60ml/4 tbsp Cointreau or orange liqueur, plus more for flaming (optional)
- brandy, for flaming (optional)
- orange segments, to decorate

1 Sift the flour, salt and sugar into a large bowl. Make a well in the centre and pour in the eggs. Beat the eggs, gradually incorporating the flour.

2 Whisk in the milk, water and orange flower water or liqueur to make a very smooth batter. Strain into a jug (pitcher) and set aside for 20–30 minutes.

3 Heat an 18–20cm/7–8in crêpe pan (preferably non-stick) over a medium heat. If the crêpe batter has thickened, add a little more water or milk to thin it. Stir the melted butter into the batter.

4 Brush the hot pan with a little extra melted butter and pour in about 30ml/2 tbsp of batter. Quickly tilt and rotate the pan to cover the base evenly with a thin layer of batter. Cook for about 1 minute, or until the top is set and the base is golden. With a metal spatula, lift the edge to check the colour, then carefully turn over the crêpe and cook for 20–30 seconds, just to set. Tip out on to a plate.

5 Continue cooking the crêpes, stirring the batter occasionally and brushing the pan with a little more melted butter as and when necessary. Place a sheet of clear film (plastic wrap) or baking parchment between each crêpe as they are stacked to prevent them from sticking. (The crêpes can be prepared ahead to this point – put them in a plastic bag and chill until ready to use.)

6 To make the sauce, melt the butter in a large frying pan over a medium-low heat, then stir in the sugar, orange and lemon rind and juice, the additional orange juice and the orange liqueur.

7 Place a crêpe in the pan browned-side down, swirling gently to coat with the sauce. Fold it in half, then in half again to form a triangle, and push to the side of the pan. Continue heating and folding the crêpes until all are warm and covered with the sauce.

8 To flame the crêpes, heat 30–45ml/2–3 tbsp each of orange liqueur and brandy in a small pan over a medium heat. Remove the pan from the heat, carefully ignite the liquid with a match then pour evenly over the crêpes. Sprinkle over the orange segments and serve immediately.

COOK'S TIP

Cointreau is the world's leading brand of orange liqueur. It is colourless and flavoured with a mixture of bitter orange peel and sweet oranges.

Creamy Lemon Rice

This is a baked rice pudding with a difference, being subtly flavoured with lemon. It is wonderful served warm or cold with fresh fruit.

SERVES FOUR

INGREDIENTS

- 15g/½oz/1 tbsp butter, cut into small pieces, plus extra for greasing
- 50g/2oz/scant ¼ cup short grain white rice
- 600ml/1 pint/2½ cups milk
- 25g/1oz/2 tbsp caster (superfine) sugar
- finely grated rind of 1 lemon
- thinly pared and shredded orange and lemon rind, to decorate

To serve

- 225g/8oz prepared fresh fruit, such as strawberries or pineapple
- 90ml/6 tbsp crème fraîche (optional)

1 Grease a 900ml/1½ pint/3¾ cup ovenproof dish. Add the rice and pour in the milk. Set aside for 30 minutes, to allow the rice to soften a little. Preheat the oven to 150°C/300°F/Gas 2.

2 Add the caster sugar, grated lemon rind and diced butter to the rice and milk and stir gently to mix. Bake the pudding for 2–2½ hours, until the top is light golden brown.

3 Decorate with shredded orange and lemon rind and serve warm or hot with the fresh fruit. Alternatively, leave the pudding to cool completely, remove and discard the skin, then chill. Fold in the crème fraîche, if using, and decorate with the shredded citrus rinds just before serving with the fruit.

VARIATION

Replace half the milk with cream.

SYRUPY BRIOCHE SLICES WITH ICE CREAM

KEEP A FEW INDIVIDUAL BRIOCHE BUNS IN THE FREEZER TO MAKE THIS SUPER FIVE-MINUTE DESSERT. FOR A SLIGHTLY TARTER TASTE, USE LEMON INSTEAD OF ORANGE RIND.

SERVES FOUR

INGREDIENTS

- butter, for greasing
- finely grated rind and juice of 1 orange, such as Navelina or blood orange
- 50g/2oz/¼ cup caster (superfine) sugar
- 90ml/6 tbsp water
- 1.5ml/¼ tsp ground cinnamon
- 4 brioche buns
- 15ml/1 tbsp icing (confectioners') sugar
- 400ml/14fl oz/1⅔ cups vanilla ice cream

1 Lightly grease a gratin dish and set aside. Put the orange rind and juice, sugar, water and cinnamon in a heavy pan. Heat gently, stirring constantly, until the sugar has dissolved, then boil rapidly, without stirring, for 2 minutes, until thickened and syrupy.

2 Remove the orange syrup from the heat and pour it into a shallow heatproof dish. Preheat the grill (broiler). Cut each brioche vertically into three thick slices. Dip one side of each slice in the hot syrup and arrange in the gratin dish, syrupy sides down. Reserve the remaining syrup. Grill (broil) the brioche until lightly toasted.

3 Using tongs, turn the brioche slices over and dust well with icing sugar. Grill for about 3 minutes more, or until they are just beginning to caramelize around the edges.

4 Transfer the hot brioche to serving plates and top with scoops of vanilla ice cream. Spoon the remaining syrup over them and serve immediately.

VARIATION
Substitute the same amount of ground cardamom for the cinnamon.

COOK'S TIP
You could also use slices of a larger brioche, rather than buns, or madeleines, sliced horizontally in half. These are traditionally flavoured with lemon or orange flower water, making them especially tasty.

Sticky Pear Pudding with Orange Cream

Cloves add a distinctively fragrant flavour to this hazelnut, pear and coffee pudding. Accompanied by a tangy orange cream, it makes a heavenly dessert.

SERVES SIX

INGREDIENTS

- 115g/4oz/½ cup unsalted (sweet) butter, softened, plus extra for greasing
- 30ml/2 tbsp ground coffee, hazelnut-flavoured if possible
- 15ml/1 tbsp near-boiling water
- 50g/2oz/⅓ cup hazelnuts, toasted and skinned (see Cook's Tip)
- 4 ripe pears
- juice of ½ orange
- 115g/4oz/generous ½ cup golden caster (superfine) sugar, plus an extra 15ml/1 tbsp for baking
- 2 eggs, beaten
- 50g/2oz/½ cup self-raising (self-rising) flour, sifted
- pinch of ground cloves
- 8 whole cloves (optional)
- 45ml/3 tbsp maple syrup
- fine strips of pared orange rind, to decorate

For the orange cream

- 300ml/½ pint/1¼ cups whipping cream
- 15ml/1 tbsp icing (confectioners') sugar, sifted
- finely grated rind of ½ orange

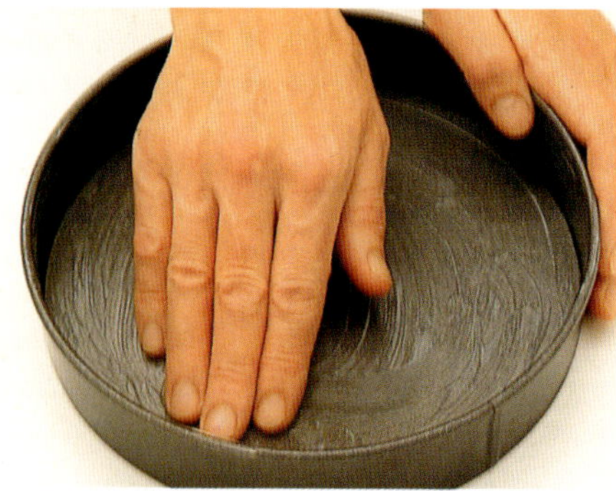

1 Preheat the oven to 180°C/350°F/Gas 4. Lightly grease a 20cm/8in loose-based sandwich tin (layer pan) with butter. Put the ground coffee in a small bowl and pour the hot water over. Leave to infuse (steep) for about 4 minutes, then strain through a fine sieve or coffee filter paper.

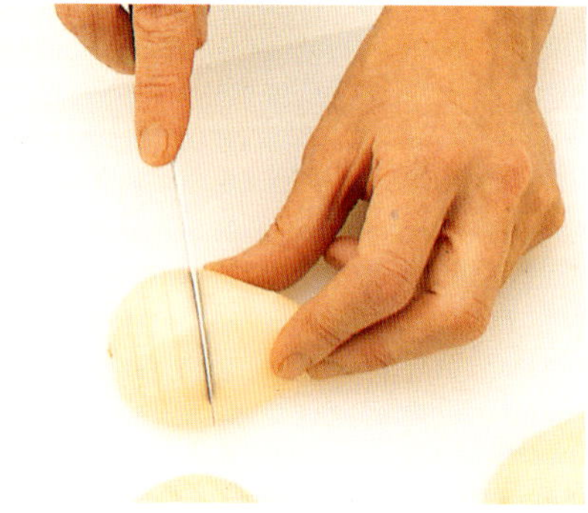

2 Grind the hazelnuts in a coffee grinder until fine. Peel, halve and core the pears. Thinly slice across the pear halves part of the way through. Brush with orange juice.

3 Beat the butter and the 115g/4oz/generous ½ cup caster sugar together in a large bowl until very light and fluffy. Gradually beat in the eggs, then fold in the flour, ground cloves, hazelnuts and coffee. Spoon the mixture into the tin and level the surface.

4 Pat the pears dry, then arrange them in the sponge mixture, flat-side down.

5 Press one whole clove into each pear half, if using, then brush the pears with 15ml/1 tbsp maple syrup.

6 Lightly sprinkle the pears with the 15ml/1 tbsp caster sugar, then bake for 45–50 minutes, or until firm.

7 While the sponge is cooking, make the orange cream. Whip the cream, icing sugar and orange rind until soft peaks form. Spoon into a serving dish and chill until needed.

8 Allow the sponge to cool for about 10 minutes in the tin, then remove and place on a serving plate. Lightly brush with the remaining maple syrup before decorating with orange rind and serving warm with the orange cream.

COOK'S TIP

To toast and skin hazelnuts, spread out in a grill (broiler) pan and toast under a hot grill for 3–4 minutes, turning them frequently until well browned. Put the nuts in a dishtowel and rub off the skins. Cool before grinding.

Coconut Rice Puddings with Grilled Oranges

Sticky rice pudding is a speciality of many South-East Asian countries. In these delightful little desserts, Thai jasmine rice is cooked with rich and creamy coconut milk, then served with golden grilled oranges.

SERVES FOUR

INGREDIENTS

- 2 oranges, such as Valencia
- 175g/6oz/scant 1 cup jasmine rice
- 400ml/14fl oz/1⅔ cup coconut milk
- 2.5ml/½ tsp freshly grated nutmeg, plus extra for sprinkling
- large pinch of salt
- 60ml/4 tbsp golden caster (superfine) sugar
- oil, for greasing
- orange peel twists, to decorate

1 Using a sharp knife, cut away the peel and pith from the oranges, then cut the flesh into rounds. Set aside.

2 Rinse and drain the rice. Place in a pan, cover with water, and bring to the boil. Cook for 5 minutes, until the grains are just beginning to soften. Place the rice in a muslin-lined (cheesecloth-lined) steamer, then make a few holes in the muslin to allow the steam to get through. Steam the rice for 15 minutes, or until tender.

3 Put the steamed rice in a heavy pan with the coconut milk, nutmeg, salt and sugar and cook over a low heat until the mixture begins to simmer. Simmer for about 5 minutes, or until the mixture is thick and creamy, stirring frequently to prevent the rice from sticking.

4 Spoon the rice mixture into four lightly oiled 175ml/6fl oz/¾ cup dariole moulds or ramekins and leave to cool.

5 Preheat the grill (broiler) to high. Line a baking tray or the grill rack with foil and place the orange slices on top. Sprinkle the oranges with a little grated nutmeg, then grill (broil) for 6 minutes, or until lightly golden, turning the slices halfway through cooking.

6 When the rice mixture is completely cold, run a knife around the edge of the moulds or ramekins and turn out the rice. Decorate with orange peel twists and serve with the warm orange slices.

DATE, FIG AND ORANGE PUDDING

WARM UP COLD WINTER DAYS WITH THIS SATISFYING STEAMED PUDDING OF RICH DRIED FRUITS AND REFRESHING ORANGE ENLIVENED WITH A DASH OF ORANGE LIQUEUR. FOR EXTRA SELF-INDULGENCE, SERVE IT WITH A LITTLE WHIPPED CREAM OR CUSTARD.

SERVES SIX

INGREDIENTS

- 2 oranges
- 115g/4oz/scant 1 cup stoned (pitted), ready-to-eat dried dates, chopped
- 115g/4oz/⅔ cup ready-to-eat dried figs, chopped
- 30ml/2 tbsp Cointreau or orange liqueur (optional)
- 175g/6oz/¾ cup unsalted (sweet) butter, plus extra for greasing
- 175g/6oz/¾ cup soft light brown sugar
- 3 eggs
- 75g/3oz/⅔ cup self-raising (self-rising) wholemeal (whole-wheat) flour
- 115g/4oz/1 cup unbleached self-raising (self-rising) flour
- 30ml/2 tbsp golden (light corn) syrup (optional)

1 Thinly pare a few pieces of rind from one orange and cut it into fine strips and reserve. Grate the rind from the remaining oranges and squeeze out the juice. Put the grated rind and juice in a pan. Add the chopped dates and figs and the orange liqueur, if using. Cook, covered, over a low heat for 8–10 minutes, or until the fruit is soft.

2 Leave the fruit mixture to cool, then transfer to a food processor or blender and process until fairly smooth. Press through a sieve to remove the fig seeds, if you like.

3 Cream the butter and sugar until pale and fluffy, then beat in the fruit purée. Beat in the eggs, then fold in the flours.

4 Grease a 1.5 litre/2½ pint/6 cup pudding bowl, and pour in the golden syrup, if using. Spoon in the pudding mixture. Cover the top with baking parchment, with a pleat folded down the centre, and then with pleated foil, and tie down with string.

5 Place the bowl in a large pan, and pour in enough water to come halfway up the sides of the bowl. Cover and steam for 2 hours. Check the water occasionally and top up if necessary. Serve decorated with the reserved orange rind.

Lemon Tartlets

These classic French tartlets are one of the most delicious desserts there is. A rich lemon curd is encased in a crisp pastry case and decorated with caramelized lemon slices.

MAKES TWELVE

INGREDIENTS
6 eggs, beaten
350g/12oz/1½ cups caster (superfine) sugar
115g/4oz/½ cup butter
grated rind and juice of 4 lemons
icing (confectioners') sugar, for dusting (optional)
175ml/6fl oz/¾ cup double (heavy) cream, to serve

For the pastry
225g/8oz/2 cups plain (all-purpose) flour
115g/4oz/½ cup chilled butter, diced
30ml/2 tbsp icing (confectioners') sugar
1 egg, beaten
5ml/1 tsp vanilla extract (essence)
15ml/1 tbsp chilled water

For the topping
2 lemons, well-scrubbed
75ml/5 tbsp apricot jam

1 Preheat the oven to 200°C/400°F/Gas 6. To make the pastry, sift the flour into a large mixing bowl. Using your fingertips, lightly rub the butter into the flour until the mixture resembles fine breadcrumbs. Add the icing sugar and stir well to mix.

2 Add the egg, vanilla essence and most of the chilled water, then work to a soft dough. Add a few more drops of water if necessary. Knead quickly and lightly, while still in the bowl, until a smooth dough forms.

3 Lightly butter 12 × 10cm/4in tartlet tins (muffin pans). Roll out the pastry on a lightly floured work surface to a thickness of 3mm/⅛in.

4 Using a 10cm/4in fluted pastry (cookie) cutter, cut out 12 rounds and press them into the tartlet tins. Prick the bases all over with a fork and then transfer the tins to a baking sheet.

5 Line the pastry cases with baking parchment and fill with baking beans. Bake the pastry cases for 10 minutes. Remove the paper and beans and set the tins aside while you make the filling.

6 Put the eggs, sugar and butter into a pan, and stir over a low heat until all the sugar has dissolved. Add the lemon rind and juice, and continue cooking, stirring constantly, until the lemon curd has thickened slightly.

7 Pour the lemon curd mixture into the pastry cases. Bake for 15 minutes, or until the curd filling is almost set.

8 Prepare the topping while the lemon tartlets are cooking. Cut the lemons into 12 slices, then cut the slices in half.

9 Push the jam through a fine sieve then transfer to a small pan and heat. Place two lemon slices in the centre of each tartlet, overlapping them if you wish. Lightly brush the top of the tartlet with the sieved jam, then put under the grill (broiler) and heat for 5 minutes, or until the top is caramelized and golden. Serve with cream, if you wish.

COOK'S TIP

If you are short of time, simply dust the top with icing (confectioners') sugar.

Lemon Meringue Pie

Crisp shortcrust is filled with a mouth-watering lemon filling and heaped with soft golden-topped meringue. This classic open tart never fails to please. Popular with adults and children, it is a classic Sunday lunch dessert.

SERVES SIX

INGREDIENTS

3 large (US extra large) egg yolks
30ml/2 tbsp caster (superfine) sugar
grated rind and juice of 1 lemon
25g/1oz/½ cup white breadcrumbs
250ml/8fl oz/1 cup milk

For the pastry

115g/4oz/1 cup plain (all-purpose) flour
pinch of salt
50g/2oz/¼ cup chilled butter
50g/2oz/¼ cup lard or white vegetable fat
15ml/1 tbsp caster (superfine) sugar
about 15ml/1 tbsp iced water

For the topping

3 large (US extra large) egg whites
115g/4oz/generous ½ cup caster (superfine) sugar

1 To make the pastry, sift the flour and salt into a bowl. Using your fingertips, rub in the butter and lard. Stir in the sugar and add enough iced water to make a soft dough.

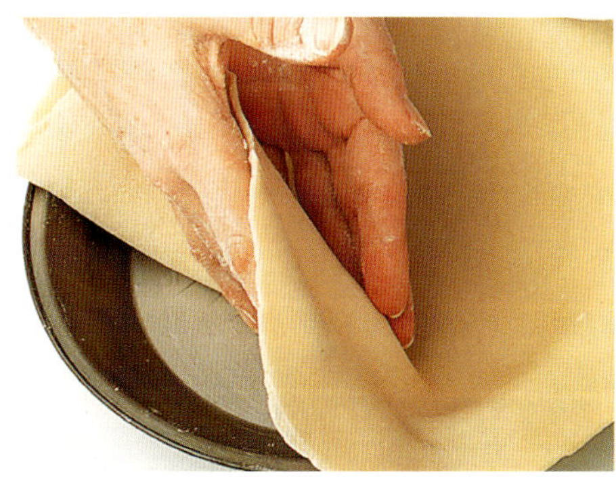

2 Roll out the pastry on a lightly floured surface and use to line a deep 21cm/8½in pie plate. Chill until required.

3 Meanwhile, make the filling. Place the egg yolks, sugar, lemon rind and juice, breadcrumbs and milk in a bowl, mix lightly and leave to soak for 1 hour.

4 Preheat the oven to 200°C/400°F/Gas 6. Beat the filling until smooth and pour into the pastry case (pie shell). Bake for 20 minutes, or until the filling has just set and the pastry is golden. Cool on a wire rack for 30 minutes, or until a skin has formed on the surface. Lower the oven temperature to 180°C/350°F/Gas 4.

5 Make the topping. Whisk the egg whites in a grease-free bowl until they form stiff peaks. Whisk in the sugar a spoonful at a time to form a glossy meringue. Spoon on top of the set lemon filling, spreading to the rim of the pastry case. Swirl the meringue slightly.

6 Bake the pie for 20–25 minutes, or until the meringue is crisp and golden brown. Leave the pie to cool on a wire rack for 10 minutes before serving.

LEMONY TREACLE TART

THIS OLD-FASHIONED FAVOURITE IS ALWAYS POPULAR AND IS ESPECIALLY LOVED BY CHILDREN, WITH ITS MELT-IN-THE-MOUTH SHORTCRUST PASTRY CASE, GLORIOUSLY STICKY LEMON AND GOLDEN SYRUP FILLING AND PRETTY TWISTED LATTICE TOPPING.

SERVES FOUR TO SIX

INGREDIENTS

- 250g/9oz/generous ¾ cup golden (light corn) syrup
- about 75g/3oz/1½ cups fresh white breadcrumbs
- grated rind of 1 lemon
- 30ml/2 tbsp lemon juice
- custard, to serve

For the pastry

- 150g/5oz/1¼ cups plain (all-purpose) flour
- 2.5ml/½ tsp salt
- 130g/4½oz/9 tbsp chilled butter, diced
- 45–60ml/3–4 tbsp chilled water

1 To make the pastry, sift the flour and salt into a mixing bowl. Rub in the butter with your fingertips until the mixture resembles fine breadcrumbs.

2 With a fork, stir in enough water to bind the dough. Gather into a ball, knead until smooth then wrap in clear film (plastic wrap) and chill for 20 minutes.

3 On a lightly floured surface, roll out the pastry to a thickness of 3mm/⅛in. Transfer to a 20cm/8in fluted flan tin (tart pan), easing it into the tin with your fingers. Trim off any overhang. Chill the pastry case (pie shell) for 20 minutes. Reserve the pastry trimmings.

4 Put a baking sheet in the oven and preheat to 200°C/400°F/Gas 6. Warm the syrup in a pan until it melts.

5 Remove the syrup from the heat and stir in the breadcrumbs and lemon rind. Leave to stand for 10 minutes, then add more breadcrumbs if the mixture is too thin. Stir in the lemon juice. Spread the mixture evenly over the pastry case.

6 Roll out the pastry trimmings and cut into 10–12 thin strips.

7 Twist the strips into spirals. Lay half of them across the filling. Arrange the remaining strips at right angles to form a lattice. Press the ends on to the rim and trim off any excess pastry. Bake for 10 minutes, then lower the temperature to 190°C/375°F/Gas 5. Bake for about 15 minutes more until golden. Serve warm with custard.

Fresh Lemon Tart

Made famous by its French title – Tarte aux Citron – this tart should be served at room temperature if the zesty lemon flavour is to be enjoyed to the full.

SERVES SIX TO EIGHT

INGREDIENTS

- 350g/12oz packet ready-made rich sweet shortcrust pastry, thawed if frozen

For the filling

- 3 eggs
- 115g/4oz/½ cup caster (superfine) sugar
- 115g/4oz/1 cup ground almonds
- 105ml/7 tbsp double (heavy) cream
- grated rind and juice of 2 lemons

For the topping

- 2 thin-skinned unwaxed lemons, thinly sliced
- 200g/7oz/scant 1 cup caster (superfine) sugar
- 105ml/7 tbsp water

COOK'S TIP

If you prefer not to candy the lemons, simply dust the tart with icing sugar.

1 Roll out the pastry and line a deep 23cm/9in fluted flan tin (tart pan). Prick the base and chill for 30 minutes.

2 Preheat the oven to 200°C/400°F/ Gas 6. Line the pastry with non-stick baking paper and baking beans and bake blind for 10 minutes. Remove the paper and beans and return the pastry case to the oven for 5 minutes more.

3 Meanwhile, make the filling. Beat the eggs, caster sugar, almonds and cream in a bowl until smooth. Beat in the lemon rind and juice. Pour the filling into the pastry case. Lower the oven temperature to 190°C/375°F/Gas 5 and bake for 20 minutes or until the filling has set and the pastry is golden.

4 Make the topping. Place the lemon slices in a pan and pour over water to cover. Simmer for 15–20 minutes or until the skins are tender, then drain.

5 Place the sugar in a saucepan and stir in the measured water. Heat gently until the sugar has dissolved, stirring constantly, then boil for 2 minutes. Add the lemon slices and cook for 10–15 minutes until the skins become shiny and candied.

6 Lift out the candied lemon slices and arrange them in rings over the top of the tart. Return the syrup to the heat and boil until it is reduced to a thick glaze. Brush this over the tart and serve warm or at room temperature.

APPLE AND ORANGE PIE

TANGY ORANGES AND MELT-IN-THE MOUTH APPLES ARE TOPPED WITH A CRISP CRUST IN THIS SATISFYING PIE THAT IS SIMPLE TO MAKE. PERFECT FOR A FAMILY DESSERT AT THE WEEKEND, IT IS ALSO SPECIAL ENOUGH FOR AN INFORMAL SUPPER PARTY.

SERVES FOUR TO SIX

INGREDIENTS
- 3 navel oranges
- 1kg/2¼lb cooking apples, peeled, cored and thickly sliced
- 30ml/2 tbsp demerara (raw) sugar
- beaten egg, to glaze
- caster (superfine) sugar, for sprinkling

For the pastry
- 275g/10oz/2½ cups plain (all-purpose) flour
- 2.5ml/½ tsp salt
- 150g/5oz/10 tbsp chilled butter, diced
- about 40ml/4 tbsp chilled water

1 To make the pastry, sift the flour and salt into a large bowl. Rub in the butter with your fingertips, until the mixture resembles fine breadcrumbs. Mix in the water and knead lightly to form a firm dough. Wrap the dough in clear film (plastic wrap) and chill for at least 30 minutes.

2 Roll out the pastry on a lightly floured work surface to a round 2cm/¾in larger than the top of a 1.2 litre/2 pint/5 cup pie dish. Cut off a narrow strip around the edge of the pastry and firmly attach it to the rim of the pie dish with a little cold water.

COOK'S TIP
Use any excess pastry to make leaves for decorating the pie, marking the veins with the blade of a knife.

3 Preheat the oven to 190°C/375°F/Gas 5. Using a sharp knife, cut a thin slice of peel and pith from both ends of each orange. Place cut side down on a plate and cut off the peel and pith in strips. Remove any bits of remaining pith. Cut out each segment leaving the membrane behind. Squeeze the remaining juice from the membrane.

4 Mix together the orange segments and juice, the apples and sugar in the pie dish. Place a pie funnel in the centre of the dish.

VARIATIONS
- Substitute 2 pink grapefruit for the oranges and double the quantity of sugar.
- Replace the cooking apples with firm pears such as Conference.

5 Dampen the pastry strip on the rim of the dish and cover with the pastry. Press the edges to the pastry strip.

6 Brush the top with beaten egg to glaze, then bake for 35 minutes, or until golden. Sprinkle the pie with caster sugar before serving.

MINCE PIES WITH ORANGE WHISKY BUTTER

MINCEMEAT GETS THE LUXURY TREATMENT WITH THE ADDITION OF GLACÉ CITRUS PEEL, CHERRIES AND WHISKY TO MAKE A MARVELLOUS FILLING FOR THESE TRADITIONAL FESTIVE PIES. SERVING THEM WITH A SPOONFUL OF WHISKY BUTTER IS PURE INDULGENCE.

MAKES TWELVE TO FIFTEEN

INGREDIENTS
- 225g/8oz/⅔ cup mincemeat
- 50g/2oz/¼ cup glacé (candied) citrus peel, chopped
- 50g/2oz/¼ cup glacé (candied) cherries, chopped
- 30ml/2 tbsp whisky
- 1 egg, beaten or a little milk
- icing (confectioners') sugar, for dusting

For the pastry
- 1 egg yolk
- 5ml/1 tsp grated orange rind
- 15ml/1 tbsp caster (superfine) sugar
- 10ml/2 tsp chilled water
- 225g/8oz/2 cups plain (all-purpose) flour
- 150g/5oz/10 tbsp butter, diced

For the orange whisky butter
- 75g/3oz/6 tbsp butter, softened
- 175g/6oz/1½ cups icing (confectioners') sugar, sifted
- 30ml/2 tbsp whisky
- 5ml/1 tsp grated orange rind

1 To make the pastry, lightly beat the egg yolk in a bowl, then add the grated orange rind, caster sugar and water and mix together. Cover and set aside. Sift the flour into a separate mixing bowl.

VARIATIONS
- Use either puff or filo pastry instead of shortcrust for a change.
- Replace the whisky in both the filling and the flavoured butter with Cointreau or brandy, if you like.

2 Using your fingertips, rub the diced butter into the flour until the mixture resembles fine breadcrumbs. Stir in the egg mixture and mix to a dough. Wrap in clear film (plastic wrap) and chill for 30 minutes.

3 Mix together the mincemeat, glacé peel and cherries, then add the whisky.

4 Roll out three-quarters of the pastry. With a fluted pastry (cookie) cutter stamp out rounds and line 12–15 patty tins (muffin pans). Re-roll the trimmings thinly and stamp out star shapes.

5 Preheat the oven to 200°C/400°F/Gas 6. Spoon a little filling into each pastry case (pie shell) and top with a star shape. Brush with a little beaten egg or milk and bake for 20–25 minutes, or until golden. Leave to cool.

6 Meanwhile, make the orange whisky butter. Place the softened butter, icing sugar, whisky and grated orange rind in a bowl and beat with a wooden spoon until light and fluffy.

7 To serve, lift off each pastry star, pipe a whirl of whisky butter on top of the filling, then replace the star. Lightly dust the mince pies with a little icing sugar.

COOK'S TIP

There is a wide range of small, shaped pastry (cookie) cutters available from kitchenware stores and special seasonal packs with a festive theme also include stars and Christmas trees. While metal cutters are usually the wiser buy, as these will be used only annually, cheaper plastic cutters are fine.

Cold and Iced Desserts

Cold desserts based on citrus fruits are a delightfully refreshing way to end a meal. In the summer, what could be more welcome than a cooling lemon sorbet or an orange granita? The recipes range from exquisitely decorated tarts to elegant cheesecakes, while rich, creamy mousses are beautifully balanced by the sharp, yet sweet flavour of citrus fruits. Oranges and lemons can be partnered with a vast array of other ingredients, from nuts to liqueurs and from ginger to cardamom, to create a flavour sensation that is bound to impress.

FRESH ORANGE TART

GRATED ORANGE RIND GIVES THIS RICH SHORTCRUST PASTRY ITS WONDERFUL COLOUR AND FLAVOUR. A CREAMY CUSTARD FILLING AND FRESH ORANGE TOPPING TURN IT INTO A SOPHISTICATED DESSERT.

SERVES NINE

INGREDIENTS
- 2 eggs, plus 2 egg yolks
- 150g/5oz/¾ cup caster (superfine) sugar
- 150ml/¼ pint/⅔ cup single (light) cream
- finely grated rind and juice of 1 orange
- 6–8 oranges
- fresh mint sprigs, to decorate

For the pastry
- 175g/6oz/1½ cups plain (all-purpose) flour
- 90g/3½oz/7 tbsp butter, diced
- 15ml/1 tbsp caster (superfine) sugar
- finely grated rind of 1 orange
- 1 egg yolk
- about 10ml/2 tsp orange juice

1 To make the pastry, sift the flour and rub in the butter. Add the sugar and orange rind and mix in. Beat the egg yolk with the orange juice, then add to the dry ingredients and mix to a dough.

2 Lightly knead the dough. Roll out and use to line a 20cm/8in square fluted tart tin (pan). Chill for 30 minutes.

3 Put a baking sheet in the oven and preheat to 200°C/400°F/Gas 6. Prick the pastry case (pie shell) all over with a fork and line with foil and baking beans. Place on the hot baking sheet and bake blind for 12 minutes. Remove the foil and beans and bake the pastry for a further 5 minutes.

4 Whisk the eggs, yolks and sugar in a bowl until foamy. Whisk in the cream, followed by the orange rind and juice. Pour into the pastry case and bake for 30–35 minutes, or until firm. Remove from the oven and leave to cool on a wire rack while still in the tin.

5 Using a sharp knife, cut a thin slice of peel and pith from both ends of each orange. Place cut-side down on a plate and cut off the peel and pith in strips. Remove any remaining pith. Cut out each segment leaving the membrane behind. Arrange the segments in rows on top of the tart. Chill until ready to serve, then carefully remove the tart from the tin and decorate with sprigs of fresh mint.

ORANGE SWEETHEART TART

STUNNING TO LOOK AT AND DELECTABLE TO EAT, THIS TART HAS A CRISP SHORTCRUST PASTRY CASE, SPREAD WITH APRICOT JAM AND FILLED WITH FRANGIPANE, THEN TOPPED WITH TANGY ORANGE SLICES.

SERVES EIGHT

INGREDIENTS
- 200g/7oz/1 cup sugar
- 250ml/8fl oz/1 cup fresh orange juice, strained
- 2 large navel oranges
- 75g/3oz/½ cup blanched almonds
- 50g/2oz/¼ cup butter
- 1 egg
- 15ml/1 tbsp plain (all-purpose) flour
- 45ml/3 tbsp apricot jam

For the pastry
- 175g/6oz/1½ cups plain (all-purpose) flour
- 2.5ml/½ tsp salt
- 75g/3oz/6 tbsp chilled butter, diced
- 45ml/3 tbsp chilled water

1 To make the pastry, sift the flour and salt into a mixing bowl. Add the butter and rub in with your fingertips until the mixture resembles fine breadcrumbs. Sprinkle over the water and mix to a dough. Knead the dough on a lightly floured surface for a few seconds until smooth. Wrap the dough in clear film (plastic wrap) and chill for at least 30 minutes.

2 After the pastry has rested, roll it out on a floured surface to a thickness of about 5mm/¼in. Use to line a 20cm/8in heart-shaped tart tin (pan). Trim the pastry edges and chill again for a further 30 minutes.

3 Preheat the oven to 200°C/400°F/Gas 6, with a baking sheet placed in the centre. Line the pastry case (pie shell) with baking parchment or foil and fill with baking beans. Bake blind for 10 minutes. Remove the parchment or foil and beans and cook for 10 minutes more, or until it is light and golden.

4 Put 150g/5oz/¾ cup of the sugar into a heavy pan and pour in the orange juice. Bring to the boil, stirring until the sugar has dissolved, then boil steadily for about 10 minutes, or until the liquid is thick and syrupy.

5 Cut the unpeeled oranges into 5mm/¼in slices. Add to the syrup. Simmer for 10 minutes, or until glazed. Transfer the slices to a wire rack to dry. When cool, cut in half. Reserve the syrup.

6 Grind the almonds finely in a food processor. With an electric mixer, cream the butter and remaining sugar until light. Beat in the egg and 30ml/2 tbsp of the orange syrup. Stir in the almonds, then add the flour.

7 Melt the jam over a low heat, then brush it evenly over the inside of the pastry case. Pour in the ground almond mixture. Bake for 20 minutes, or until set. Leave to cool in the tin.

8 Starting at the top of the heart shape and working down to the point, arrange the orange slices on top of the tart in an overlapping pattern, cutting them to fit. Boil the remaining syrup until thick and brush on top to glaze. Leave to cool.

Lemon Cheese Mousse with Brandy Snap Baskets

A light cheesecake-style lemon mousse with a hint of ginger fills these dainty brandy snap baskets. Assemble the individual desserts at the last minute, so that they stay crisp.

MAKES SIX TO EIGHT

INGREDIENTS

45ml/3 tbsp water
10ml/2 tsp powdered gelatine
250g/9oz/generous 1 cup curd (farmer's) cheese
150ml/¼ pint/⅔ cup natural (plain) yogurt
juice of 2 lemons
30ml/2 tbsp clear honey, or to taste
15ml/1 tbsp grated crystallized (candied) ginger
2 egg whites
mint sprigs, to decorate
selection of soft fruit, to serve

For the baskets

50g/2oz/¼ cup butter
30ml/2 tbsp golden (light corn) syrup
50g/2oz/¼ cup granulated sugar
grated rind of 2 lemons
50g/2oz/½ cup plain (all-purpose) flour
1 orange, for shaping the baskets

1 To make the baskets, preheat the oven to 190°C/375°F/Gas 5. Line a large baking sheet with baking parchment. Melt the butter, syrup and sugar in a pan, then remove from the heat and stir in half the lemon rind and all the flour. Beat until smooth.

COOK'S TIP

To save time, you can use bought brandy snaps to make the baskets. Simply warm them in a very low oven until they uncurl, then shape them into baskets over the orange.

2 Put about 30ml/2 tbsp of mixture on the prepared baking sheet. Using a metal spatula, spread out the mixture to a 13cm/5in circle. Add a second circle, some distance from the first to allow room for spreading. Bake for 5–7 minutes, or until the biscuits (cookies) have spread out and are lacy and light golden brown.

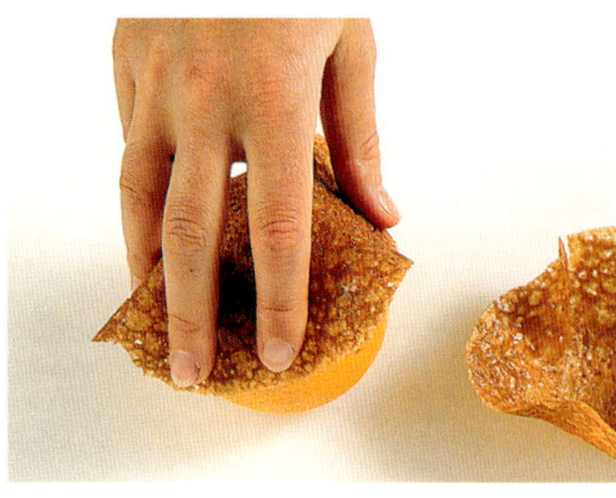

3 Allow the biscuits to cool for about 1 minute so that they firm slightly, but do not harden, then lift each in turn off the baking sheet using a metal spatula. Gently press the biscuits over the orange, carefully fluting the edges, and protecting your hands with a clean dishtowel, if necessary.

4 Remove the baskets when shaped and leave them to cool on a wire rack. Repeat the entire cooking and shaping process with the remaining mixture until you have used it all. If the brandy snaps harden before you have a chance to shape them, simply pop them back in the oven briefly to soften.

5 Make the mousse. Pour the water into a bowl and sprinkle the gelatine over the surface. Leave until spongy, then dissolve over simmering water.

6 Blend the cheese, yogurt, remaining lemon rind, lemon juice, honey and ginger in a food processor. Add the gelatine, blend briefly and pour into a bowl. Chill until just beginning to set.

7 Whisk the egg whites to soft peaks and fold into the mousse. Spoon into the baskets and serve, decorated with mint sprigs, accompanied by soft fruit.

LEMON AND LIME CHEESECAKE

Here is a lemon and lime cheesecake that is a cut above the rest. Topped with a zesty lime syrup, it is truly a citrus sensation.

SERVES EIGHT

INGREDIENTS

- 150g/5oz digestive biscuits (graham crackers)
- 40g/1½oz/3 tbsp butter

For the topping

- grated rind and juice of 2 lemons
- 10ml/2 tsp powdered gelatine
- 250g/9oz/generous 1 cup ricotta cheese
- 75g/3oz/⅓ cup caster (superfine) sugar
- 150ml/¼ pint/⅔ cup double (heavy) cream
- 2 eggs, separated

For the lime syrup

- finely pared rind and juice of 3 limes
- 75g/3oz/⅓ cup caster (superfine) sugar
- 5ml/1 tsp arrowroot mixed with 30ml/2 tbsp water
- a little green food colouring (optional)

1 Grease a 20cm/8in round springform cake tin (pan). Process the biscuits in a food processor to fine crumbs. Melt the butter in a pan, then stir in the crumbs. Spoon into the cake tin, press the crumbs down well in an even layer, then cover and chill.

2 Make the topping. Put the lemon rind and juice in a pan and sprinkle over the gelatine. Leave for 5 minutes, then heat until the gelatine has dissolved. Set aside to cool slightly. Beat the ricotta and sugar. Stir in the cream and egg yolks. Whisk in the gelatine mixture.

3 Whisk the egg whites in a grease-free bowl until they form soft peaks. Fold them into the cheese mixture. Spoon on to the biscuit base, level the surface and chill for 2–3 hours.

4 Meanwhile, make the lime syrup. Cut the lime rind into very fine shreds. Place the lime shreds, juice and caster sugar in a small pan. Bring to the boil, stirring constantly, then boil the syrup for 5 minutes, without stirring.

5 Stir in the arrowroot mixture and continue to stir over a low heat until the syrup boils again and thickens slightly. Tint the syrup pale green with a little food colouring, if you like. Cool, then chill until required.

6 Spoon the lime syrup over the set cheesecake. Remove from the tin and cut into slices to serve.

LEMON CHEESECAKE WITH FOREST FRUITS

NO COLLECTION OF CITRUS RECIPES WOULD BE COMPLETE WITHOUT A CREAMY LEMON CHEESECAKE. THIS ONE HAS A LIGHT CORNFLAKE BASE AND IS SERVED TOPPED WITH LUSCIOUS FOREST FRUITS.

SERVES EIGHT

INGREDIENTS
- 50g/2oz/¼ cup unsalted (sweet) butter
- 25g/1oz/2 tbsp light soft brown sugar
- 45ml/3 tbsp golden (light corn) syrup
- 115g/4oz/generous 1 cup cornflakes
- 11g/¼oz sachet powdered gelatine
- 225g/8oz/1 cup soft cheese
- 150g/5oz/generous ½ cup Greek (US strained plain) yogurt
- 150ml/¼ pint/⅔ cup single (light) cream
- finely grated rind and juice of 2 lemons
- 75g/3oz/6 tbsp caster (superfine) sugar
- 2 eggs, separated
- 225g/8oz/2 cups mixed, prepared fresh forest fruits, such as blackberries, raspberries and redcurrants, to decorate
- icing (confectioners') sugar, for dusting

1 Place the butter, brown sugar and syrup in a pan and heat over a low heat, stirring, until the mixture has melted and is well blended. Remove from the heat and stir in the cornflakes.

2 Press the mixture over the base of a deep 20cm/8in loose-based round cake tin (pan). Chill for 30 minutes.

VARIATION
Use unsweetened puffed rice cereal or rice crispies in place of the cornflakes.

3 Sprinkle the gelatine over 45ml/3 tbsp water in a bowl and leave to soak for a few minutes. Place the bowl over a pan of simmering water and stir until the gelatine has dissolved. Place the cheese, yogurt, cream, lemon rind and juice, caster sugar and egg yolks in a large bowl and beat until smooth and thoroughly mixed.

4 Add the hot gelatine to the cheese and lemon mixture and beat well.

5 Whisk the egg whites until stiff, then fold into the cheese mixture.

6 Pour the cheese mixture over the cornflake base and level the surface. Chill for 4–5 hours, or until the filling has set.

7 Carefully remove the cheesecake from the tin and place on a serving plate. Decorate with the mixed fresh fruits, dust with icing sugar and serve.

COLD LEMON SOUFFLÉ WITH ALMONDS

TERRIFIC TO LOOK AT YET EASY TO MAKE, THIS REFRESHING LEMON SOUFFLÉ IS LIGHT AND MOUTHWATERING, IDEAL FOR THE END OF ANY MEAL.

SERVES SIX

INGREDIENTS

- oil, for greasing
- grated rind and juice of 3 large lemons
- 5 large (US extra large) eggs, separated
- 115g/4oz/generous ½ cup caster (superfine) sugar
- 25ml/1½ tbsp powdered gelatine
- 450ml/¾ pint/scant 2 cups double (heavy) cream

For the topping

- 75g/3oz/¾ cup flaked (sliced) almonds
- 75g/3oz/¾ cup icing (confectioners') sugar

1 Cut a strip of baking parchment long enough to fit around a 900ml/1½ pint/ 3¾ cup soufflé dish and wide enough to extend 7.5cm/3in above the rim. Fit the strip around the dish, tape, then tie it around the top of the dish with string.

2 Using a pastry brush, lightly coat the inside of the paper collar with oil.

3 Put the lemon rind and egg yolks in a bowl. Add 75g/3oz/6 tbsp of the caster sugar and whisk thoroughly until the mixture becomes light and creamy.

COOK'S TIP

To dissolve the gelatine more quickly, heat the lemon juice and gelatine in a microwave, on full power, in 30-second bursts, stirring between each burst, until it is fully dissolved.

4 Place the lemon juice in a heatproof bowl and sprinkle over the gelatine. Set aside for 5 minutes, then place the bowl in a pan of simmering water. Heat, stirring occasionally, until the gelatine has dissolved. Cool slightly, then stir into the egg yolk mixture.

5 Whip the cream to soft peaks. Fold into the egg yolk mixture and set aside.

6 Whisk the egg whites to stiff peaks, then whisk in the remaining caster sugar until stiff. Fold the whites into the yolk mixture. Pour into the dish, smooth the surface and chill for 4–5 hours.

7 To make the topping, brush a baking sheet lightly with oil. Preheat the grill (broiler). Sprinkle the almonds over the baking sheet and sift the icing sugar over them. Grill (broil) until the nuts have turned a rich golden colour and the sugar has caramelized.

8 Leave to cool, then remove the almond mixture from the tray with a metal spatula and break it into pieces.

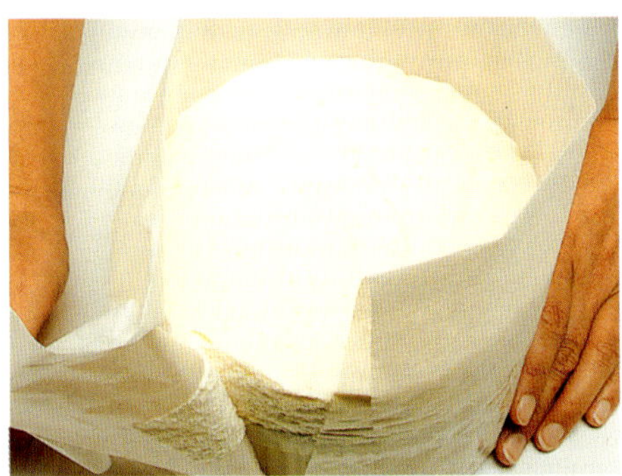

9 When the soufflé has set, peel off the paper. If the paper does not come away easily, hold the flat blade of a knife against the set soufflé as you peel the paper away – this will help the soufflé keep its shape. Sprinkle the caramelized almonds on top of the lemon soufflé, and serve immediately.

VARIATION

This soufflé is wonderfully refreshing when served semi-frozen. Place the undecorated, set soufflé in the freezer for about an hour. Just before serving, remove from the freezer and decorate with the caramelized almonds.

LEMON COEUR À LA CRÈME WITH ORANGES

These charming heart-shaped, creamy sweet cheeses are a traditional French dessert. Surrounded by oranges in a Cointreau sauce, they make a delightful and refreshing finale to a rich main course.

SERVES FOUR

INGREDIENTS

- 225g/8oz/1 cup cottage cheese
- 250g/9oz/generous 1 cup mascarpone cheese
- 50g/2oz/¼ cup caster (superfine) sugar
- grated rind and juice of 1 lemon
- spirals of orange rind, to decorate (see Cook's Tip)

For the Cointreau oranges

- 5 oranges
- 10ml/2 tsp cornflour (cornstarch)
- 15ml/1 tbsp icing (confectioners') sugar
- 60ml/4 tbsp Cointreau

1 Put the cottage cheese in a food processor or blender and process until smooth. Add the mascarpone, caster sugar, the grated lemon rind and lemon juice and process briefly to thoroughly mix the ingredients.

2 Line four *coeur à la crème* moulds with muslin (cheesecloth), then divide the mixture among them. Level the surface of each, then place the moulds on a plate to catch any liquid that drains from the cheese. Cover with clear film (plastic wrap) and chill overnight.

3 To make the Cointreau oranges, squeeze the juice from two oranges and pour into a measuring jug (cup). Make the juice up to 250ml/8fl oz/1 cup with water, if necessary, then pour the liquid into a small pan. Blend a little of the mixture with the cornflour and then add to the pan with the icing sugar. Gently heat, stirring until thickened.

4 Cut a thin slice of peel and pith from both ends of the remaining oranges. Place cut side down on a plate and cut off the peel and pith. Remove any bits of remaining pith. Cut out each segment leaving the membrane behind. Add the segments to the pan, stir to coat well then set aside. Cool, then stir in the Cointreau. Cover and chill overnight.

5 Turn the moulds out on to plates and surround the cheeses with the oranges. Decorate with spirals of orange rind and serve immediately.

COOK'S TIP

To make fine orange rind spirals, use a cannelle knife (zester) and pare long, narrow strips of rind. Twist tightly around cocktail sticks (toothpicks) so that they curl into corkscrews. Slide the sticks out.

STRAWBERRY AND LEMON CURD GÂTEAU

In this frozen dessert, strawberry ice cream and a layer of lemon curd cream top a light sponge cake. Fresh, juicy strawberries complete this gâteau making it perfect for summer entertaining or a family celebration.

SERVES EIGHT

INGREDIENTS

- 115g/4oz/½ cup unsalted (sweet) butter, softened, plus extra for greasing
- 115g/4oz/generous ½ cup caster (superfine) sugar
- 2 eggs
- 115g/4oz/1 cup self-raising (self-rising) flour
- 2.5ml/½ tsp baking powder
- 500ml/17fl oz/2¼ cups strawberry ice cream, softened
- 300ml/½ pint/1¼ cups double (heavy) cream
- 200g/7oz/¾ cup good quality lemon curd
- 30ml/2 tbsp lemon juice

For the topping

- 500g/1¼lb/5 cups fresh strawberries, hulled
- 25g/1oz/2 tbsp caster (superfine) sugar
- 45ml/3 tbsp Cointreau or other orange-flavoured liqueur

1 Preheat the oven to 180°C/350°F/Gas 4. Grease and line a 23cm/9in springform cake tin (pan) with baking parchment. In a large mixing bowl, beat the butter with the sugar, eggs, flour and baking powder until creamy, pale and fluffy.

2 Spoon the mixture into the prepared tin and bake for about 20 minutes, or until just firm. Leave to cool for about 5 minutes, then turn the cake out on a wire rack. Leave to cool completely.

3 Wash and thoroughly dry the cake tin, ready to use again. Line the sides of the clean cake tin with a strip of baking parchment.

4 Using a sharp knife, carefully slice off the top of the cake where it has formed a crust. Save this for another purpose. Fit the cake in the tin, cut side down. Freeze for 10 minutes, then spread the ice cream evenly over the sponge and freeze until firm.

5 Pour the cream into a bowl, whip it until it forms soft peaks, then gently fold in the lemon curd and lemon juice. Spoon the mixture over the strawberry ice cream. Cover with clear film (plastic wrap) and freeze overnight.

6 About 45 minutes before you intend to serve the dessert, make the topping. Cut half the strawberries into thin slices. Put the remainder in a food processor or blender and add the caster sugar and Cointreau or other liqueur. Process the mixture to make a smooth purée.

7 Arrange the sliced strawberries over the top of the frozen gâteau and place in the refrigerator to soften. Serve with the sauce spooned over the top.

VARIATION

Use raspberry ice cream and fresh raspberries instead of strawberry ice cream and strawberries.

LEMON SORBET

THIS HAS TO BE THE CLASSIC SORBET. REFRESHINGLY TANGY AND YET DELICIOUSLY SMOOTH, IT IS WELCOME ALL YEAR ROUND, NOT JUST DURING THE WARMER MONTHS.

SERVES SIX

INGREDIENTS
- 200g/7oz/1 cup caster (superfine) sugar
- 300ml/½ pint/1¼ cups water
- 4 lemons, well scrubbed
- 1 egg white
- sugared lemon rind, to decorate (see Cook's Tip)

1 Put the sugar and water into a pan and bring to the boil over a low heat, stirring occasionally until the sugar has just dissolved.

2 Using a vegetable peeler pare the rind thinly from two of the lemons so that it falls straight into the pan.

COOK'S TIP
To make sugared lemon rind, thinly pare the rind with a zester. Dust with a little caster (superfine) sugar.

VARIATION
Substitute four oranges for the lemons or use two oranges and two lemons for a mixed citrus sorbet.

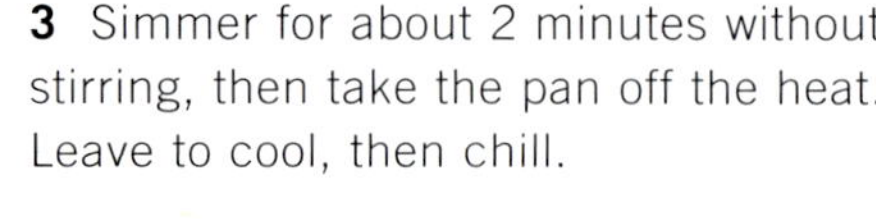

3 Simmer for about 2 minutes without stirring, then take the pan off the heat. Leave to cool, then chill.

4 Squeeze the juice from all the lemons and add it to the syrup. Strain the syrup into a shallow freezerproof container, reserving the rind. Freeze the mixture for 4 hours, until it is mushy. If you are using an ice cream maker, strain the syrup and lemon juice and then churn the mixture until thick.

5 Scoop the sorbet (sherbet) into a food processor and beat it until smooth. Lightly whisk the egg white with a fork until it is just frothy. Spoon the sorbet back into the container, beat in the egg white and return the mixture to the freezer for 4 hours. If you are using an ice cream maker, add the egg white to the mixture and continue to churn for 10–15 minutes, or until the sorbet is firm enough to scoop.

6 Scoop the sorbet into bowls or sundae glasses, decorate with sugared lemon rind and serve.

LEMON AND CARDAMOM ICE CREAM

THE CLASSIC PARTNERSHIP OF LEMON AND CARDAMOM GIVES THIS RICH ICE CREAM A LOVELY CLEAN TANG. IT IS THE PERFECT CHOICE FOR SERVING AFTER A SPICY MAIN COURSE.

SERVES SIX

INGREDIENTS
- 15ml/1 tbsp cardamom pods
- 4 egg yolks
- 115g/4oz/generous ½ cup caster (superfine) sugar
- 10ml/2 tsp cornflour (cornstarch)
- grated rind and juice of 3 lemons
- 300ml/½ pint/1¼ cups milk
- 300ml/½ pint/1¼ cups whipping cream
- fresh lemon balm sprigs and icing (confectioners') sugar, to decorate

1 Put the cardamom pods in a mortar and crush them with a pestle to release the seeds. Pick out and discard the shells, then grind the seeds to break them up slightly.

2 Put the egg yolks, sugar, cornflour, lemon rind and juice in a bowl. Add the cardamom seeds and whisk well.

COOK'S TIPS
- Lemon balm, which has a pronounced lemony flavour and fragrance, is an easy herb to grow. The leaves are best picked before the flowering period, when they are at their most fragrant. Lemon balm does not dry successfully.
- Transfer the ice cream from the freezer to the refrigerator about 30 minutes before serving so that it softens slightly.
- Always buy cardamom whole, in the pod, and grind the seeds as you require them. Once ground, cardamom quickly loses its flavour.

3 Bring the milk to the boil in a pan, then pour over the egg yolk mixture, stirring well. Return the mixture to the pan and cook over a very low heat, stirring constantly until thickened.

4 Pour the custard into a bowl, cover the surface with baking parchment and leave to cool. Chill until very cold.

5 Whip the cream lightly and fold into the custard. Pour into a container and freeze for 3–4 hours, beating twice to break up the ice crystals. If you are using an ice cream maker, whisk the cream lightly into the custard and churn the mixture until it holds its shape. Scoop into glasses and decorate with the lemon balm and icing sugar.

Tequila and Orange Granita

Full of citrus flavour enlivened with tequila, this distinctive Mexican-style granita will have guests clamouring for more. Serve simply with wedges of citrus fruit or spoon over a little grenadine for extra colour.

SERVES SIX

INGREDIENTS

- 115g/4oz/generous ½ cup caster (superfine) sugar
- 300ml/½ pint/1¼ cups water
- 6 oranges, well scrubbed
- 90ml/6 tbsp tequila
- orange and lime wedges, to serve

VARIATION

If you don't have any tequila, make the granita with vodka, Cointreau, or even white rum. Do not be tempted to add more than the recommended amount; too much alcohol will prevent the granita from freezing.

1 Put the sugar and water into a pan. Using a vegetable peeler, thinly pare the rind from three of the oranges, letting it fall into the pan. Bring to the boil, stirring to dissolve the sugar. Pour the syrup into a bowl, cool, then chill.

2 Strain the syrup into a shallow, freezer-proof plastic container. Squeeze all the oranges, strain the juice into the syrup and then stir in the tequila. Check that the mixture is no more than about 2.5cm/1in deep; transfer to a larger container if needed.

3 Cover and freeze for about 2 hours, or until the mixture around the sides of the container is mushy. Mash well with a fork to break up the ice crystals and return the granita to the freezer.

4 Freeze for a further 2 hours, mashing the mixture with a fork to break up the crystals every 30 minutes or so, or until the granita has a slushy consistency. Scoop it into dishes and serve with orange and lime wedges.

COOK'S TIP

There are two types of tequila – white and golden. The latter is aged, has a more subtle flavour and, therefore, is a little more expensive. Both types are quite suitable for this granita.

ORANGE FLOWER WATER ICE CREAM

DELICATELY PERFUMED WITH ORANGE FLOWER WATER AND A LITTLE GRATED ORANGE RIND, THIS NUTTY, LIGHTLY SWEETENED ICE CREAM TAKES ITS INSPIRATION FROM DESSERTS THAT ARE POPULAR THROUGHOUT THE MIDDLE EAST.

SERVES FOUR TO SIX

INGREDIENTS

- 4 egg yolks
- 75g/3oz/6 tbsp caster (superfine) sugar
- 5ml/1 tsp cornflour (cornstarch)
- 300ml/½ pint/1¼ cups milk
- 300ml/½ pint/1¼ cups whipping cream
- 150g/5oz/1¼ cups cashew nuts, finely chopped
- 15ml/1 tbsp orange flower water
- grated rind of ½ orange, plus spirals of orange rind, to decorate

1 Whisk the egg yolks, caster sugar and cornflour in a bowl until thick. Pour the milk into a pan and bring it to the boil. Whisk it into the egg yolk mixture.

2 Return to the pan and cook over a low heat, stirring constantly, until very smooth. Pour back into the bowl. Leave to cool, then chill.

3 Heat the cream in a pan. When it boils, stir in the chopped cashew nuts. Leave to cool.

4 Stir the orange flower water and grated orange rind into the chilled custard. Process the cashew nut cream in a food processor or blender until it forms a fine paste, then stir it into the custard mixture.

5 Pour the mixture into a plastic tub or similar freezerproof container and freeze for 6 hours, beating twice with a fork or whisking briefly with an electric mixer to break up the ice crystals until smooth. If you are using an ice cream maker, churn the mixture until it is firm enough to scoop.

6 To serve, scoop into dishes and decorate with orange rind curls.

LEMON SORBET CUPS WITH SUMMER FRUITS

IN THIS STUNNING DESSERT, LEMON SORBET IS MOULDED INTO CUP SHAPES TO MAKE PRETTY CONTAINERS FOR A SELECTION OF SUMMER FRUITS. THEY ARE IDEAL FOR ENTERTAINING, AS YOU CAN PREPARE THE CUPS MANY DAYS IN ADVANCE AND JUST FILL THEM WHEN READY TO SERVE.

SERVES SIX

INGREDIENTS

- 500ml/17fl oz/2¼ cups lemon sorbet (sherbet)
- 225g/8oz/2 cups small strawberries
- 150g/5oz/scant 1 cup raspberries
- 75g/3oz/¾ cup redcurrants, blackcurrants or white currants
- 15ml/1 tbsp caster (superfine) sugar
- 45ml/3 tbsp Cointreau or other orange-flavoured liqueur

1 Put six 150ml/¼ pint/⅔ cup metal moulds in the freezer for 15 minutes to chill. At the same time, remove the sorbet from the freezer to soften slightly.

2 Using a teaspoon, pack the sorbet into the moulds, building up a layer about 1cm/½in thick around the base and sides, and leaving a deep cavity in the centre. Return each mould to the freezer when it is lined.

3 Cut the strawberries in half and place in a bowl with the raspberries and red, black or white currants. Add the sugar and liqueur and toss the ingredients together lightly. Cover and chill for at least 2 hours.

4 Once the sorbet in the moulds has frozen completely, loosen the edges with a knife, then dip in a bowl of very hot water for 2 seconds. Invert the sorbet cups on a small tray, using a fork to twist and loosen the cups if necessary.

5 If you need to, dip the moulds very briefly in the hot water again. Turn out the sorbet cups and turn them over so they are ready to fill. Return them to the freezer until required.

6 To serve, place the cups on serving plates and fill with the fruits, spooning over any juices.

COOK'S TIP

When lining the moulds with the sorbet (sherbet), wrap your hand in a dishtowel.

Iced Oranges

These tangy little ices, served in the orange shells, were originally sold in beach cafés in the south of France. They are pretty and easy to eat – equally at home at a dinner party, or prepared as a picnic treat to store in the cold box.

SERVES EIGHT

INGREDIENTS
- about 150g/5oz/¾ cup granulated sugar
- juice of 1 lemon
- 14 medium oranges
- fresh orange juice, if needed
- 8 fresh bay leaves, to decorate

1 Put the sugar in a heavy pan. Add half the lemon juice, and 120ml/4fl oz/½ cup water. Cook over a low heat, stirring constantly until the sugar has dissolved. Bring to the boil, and boil, without stirring, for 2–3 minutes, or until the syrup is clear. Remove the pan from the heat and leave to cool.

2 Slice the tops off eight of the oranges, to make "hats". Scoop out the flesh of the oranges, and reserve. Put the empty orange shells and hats on a tray and place in the freezer.

3 Grate the rind of the remaining oranges and add to the cooled syrup.

4 Squeeze the juice from the oranges, and from the reserved flesh. There should be about 750ml/1¼ pint/3 cups. Squeeze another orange or add bought orange juice, if necessary to make up to the required quantity.

5 Stir the orange juice and remaining lemon juice, with 90ml/6 tbsp water into the syrup. Pour into a freezerproof container, cover and freeze for 3 hours.

6 Turn the mixture into a large bowl, and quickly whisk with a small balloon whisk or electric mixer to break down the ice crystals.

7 Return the citrus mixture to the freezerproof container and freeze for about 4 hours more, until just firm, but not solid.

8 Pack the mixture into the orange shells, mounding it up, and set the hats on top. Freeze until ready to serve. Just before serving, push a skewer into the tops of the hats and push in a bay leaf for decoration.

COOK'S TIP
Use crumpled kitchen paper between the shells to keep them upright.

Ruby Orange Sherbet in Ginger Baskets

This superb frozen dessert is perfect for people without ice cream makers who cannot be bothered with the freezing and stirring that home-made ices normally require. It is also ideal for serving at a special dinner party as both the sherbet and ginger baskets can be made in advance and the dessert simply assembled between courses.

SERVES SIX

INGREDIENTS
- grated rind and juice of 2 blood oranges
- 175g/6oz/1½ cups icing (confectioners') sugar
- 300ml/½ pint/1¼ cups double (heavy) cream
- 200g/7oz/scant 1 cup Greek (US strained plain) yogurt
- blood orange segments, to decorate (optional)

For the ginger baskets
- 25g/1oz/2 tbsp unsalted (sweet) butter
- 15ml/1 tbsp golden (light corn) syrup
- 30ml/2 tbsp caster (superfine) sugar
- 1.5ml/¼ tsp ground ginger
- 15ml/1 tbsp finely chopped mixed citrus peel
- 15ml/1 tbsp plain (all-purpose) flour
- butter, for greasing

1 Place the orange rind and juice in a bowl. Sift the icing sugar over the top and set aside for 30 minutes, then stir until smooth.

COOK'S TIP
When making the ginger baskets it is essential to work quickly. Have the greased tins (pans) or cups ready before you start. If the biscuits (cookies) cool and firm up before you have time to drape them all, return them to the oven for a few seconds to soften them again.

2 Whisk the double cream in a large bowl until the mixture forms soft peaks, then fold in the yogurt.

3 Gently stir in the orange juice mixture, then pour into a freezerproof container. Cover and freeze for about 3 hours, until firm.

4 Make the baskets. Preheat the oven to 180°C/350°F/Gas 4. Place the butter, syrup and sugar in a heavy pan and heat gently until melted.

5 Add the ground ginger, mixed citrus peel and flour and stir until the mixture is smooth.

VARIATION
For a simpler dish, you can also roll the freshly cooked ginger biscuits (cookies) around the well-greased handles of wooden spoons, in the same way as you would when making brandy snaps. Scoop the sherbet into glasses and serve it accompanied by the ginger biscuits.

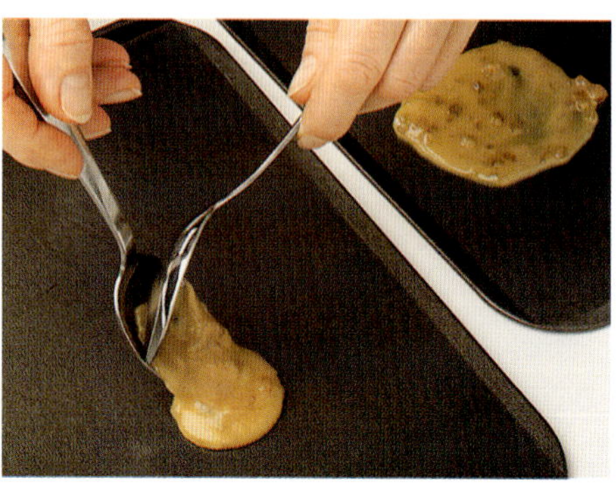

6 Lightly grease two baking sheets. Using about 10ml/2 tsp of the mixture at a time, drop three portions of the ginger dough on to each baking sheet, spacing them well apart. Spread each one to a 5cm/2in circle, then bake for 12–14 minutes, or until the biscuits (cookies) are dark golden in colour.

7 Remove the biscuits from the oven and leave to stand on the baking sheets for 1 minute to firm slightly. Carefully lift them off with a metal spatula and drape over six well-greased mini pudding tins (pans) or upturned cups; flatten the top of the biscuits (which will become the base) and flute the edges to form a decorative basket shape.

8 Once they are cool and set, lift the baskets off the tins or cups and place on individual dessert plates. Arrange small scoops of the frozen orange sherbet in each basket. Decorate each portion with a few orange segments, if you like.

Cakes, Bakes and Breads

From rich, moist cakes to speciality breads, oranges and lemons play a starring role. The fruits themselves can be used in a variety of ways, to make tangy syrups, creamy fillings and attractive decorations for that all-important finishing touch. An indulgent, lemon and lime flavoured sponge cake or a fabulous chocolate and orange marmalade teabread is sure to prove completely irresistible, and any one of these deliciously tangy recipes would make a wonderful treat for a well-earned break in a busy day.

Rich Lemon Poppy-seed Cake

The classic combination of poppy seeds and lemon is used for this tangy cake, which has a delicious lemon curd and fromage frais filling.

SERVES EIGHT

INGREDIENTS

- 350g/12oz/1½ cups unsalted (sweet) butter, plus extra for greasing
- 350g/12oz/1¾ cups golden caster (superfine) sugar
- 45ml/3 tbsp poppy seeds
- 20ml/4 tsp grated lemon rind
- 70ml/4½ tbsp lemon curd
- 6 eggs, separated
- 120ml/4fl oz/½ cup semi-skimmed (low-fat) milk
- 350g/12oz/3 cups self-raising (self-rising) flour
- icing (confectioners') sugar, to decorate

For the filling

- 150g/5oz/½ cup luxury lemon curd
- 150ml/¼ pint/⅔ cup fromage frais

1 Butter and lightly flour two 23cm/9in springform cake tins (pans). Preheat the oven to 180°C/350°F/Gas 4.

2 Cream together the butter and caster sugar in a mixing bowl with a wooden spoon until light and fluffy. Alternatively, you can use a hand-held electric whisk.

3 Add the poppy seeds, lemon rind, lemon curd and egg yolks and beat well, then add the milk and mix well. Gently fold in the flour until combined.

4 Whisk the egg whites until they form soft peaks. Fold the egg whites into the cake mixture until just combined. Divide the cake mixture evenly between the prepared tins.

5 Bake for 40–45 minutes, or until a skewer inserted into the centre of the cakes comes out clean.

6 Leave the cakes to cool in the tins for about 5 minutes, then remove from the tins and leave to cool completely on wire racks.

7 To finish, place one cake upside down on a serving plate. Spread evenly with the lemon curd for the filling and spoon the fromage frais generously over the lemon curd. Put the second cake on top, right way up, press down gently, then dust with a light coating of icing sugar before serving.

MADEIRA CAKE WITH LEMON SYRUP

THIS SUGAR-CRUSTED CAKE IS SOAKED IN A LEMON SYRUP, SO THAT IT STAYS MOIST AND IS INFUSED WITH A TANGY CITRUS FLAVOUR. SERVE IT SLICED WITH MORNING COFFEE OR AFTERNOON TEA.

SERVES TEN

INGREDIENTS

- 250g/9oz/1 cup butter, softened, plus extra for greasing
- 225g/8oz/generous 1 cup caster (superfine) sugar
- 5 eggs
- 275g/10oz/2½ cups plain (all-purpose) flour, sifted
- 30ml/2 tbsp baking powder
- salt

For the sugar crust

- 60ml/4 tbsp lemon juice
- 15ml/1 tbsp golden (light corn) syrup
- 30ml/2 tbsp sugar

1 Preheat the oven to 180°C/350°F/Gas 4. Lightly grease a 1kg/2¼lb loaf tin (pan). Beat the butter and sugar until light and creamy, then gradually beat in the eggs.

2 Mix the sifted flour, baking powder and salt, and gently fold into the egg mixture. Spoon into the prepared tin, level the top and bake for 1¼ hours, or until a skewer pushed into the middle comes out clean.

3 Remove the cake from the oven and, while still warm and in the tin, use a skewer to pierce it several times.

4 To make the sugar crust, warm the lemon juice and syrup, add the sugar and immediately spoon over the cake. Leave to cool, then chill for several hours.

Marmalade Teabread

Orange marmalade and cinnamon give this moist teabread a deliciously warm flavour. It is sure to be popular with the whole family.

SERVES EIGHT

INGREDIENTS

- 90g/3½oz/7 tbsp butter or margarine, plus extra for greasing
- 200g/7oz/1¾ cups plain (all-purpose) flour
- 5ml/1 tsp baking powder
- 6.25ml/1¼ tsp ground cinnamon
- 50g/2oz/4 tbsp soft light brown sugar
- 60ml/4 tbsp chunky orange marmalade
- 1 egg, beaten
- about 45ml/3 tbsp milk
- 50g/2oz/½ cup icing (confectioners') sugar
- about 15ml/1 tbsp warm water
- thinly pared and shredded orange and lemon rind, to decorate

1 Preheat the oven to 160°C/325°F/ Gas 3. Butter a 900ml/1½ pint/3¾ cup loaf tin (pan), then line the base with greased baking parchment.

2 Sift the flour, baking powder and cinnamon into a large bowl, then rub in the butter with your fingertips until the mixture resembles fine breadcrumbs. Stir in the sugar.

3 Mix together the marmalade, egg and most of the milk, then stir into the flour mixture to make a soft dropping (pourable) consistency, adding a little more milk if necessary.

4 Transfer the mixture to the tin and bake for about 1¼ hours until firm to the touch. Leave the cake to cool for 5 minutes, then turn on to a wire rack.

5 Carefully peel off the lining paper and leave the cake to cool completely.

6 When the cake is cold, make the icing. Sift the icing sugar into a bowl and mix in the water a little at a time to make a thick glaze. Drizzle the icing over the top of the cake and decorate with the orange and lemon rinds.

ORANGE MARMALADE CHOCOLATE LOAF

DO NOT BE ALARMED AT THE AMOUNT OF CREAM IN THIS RECIPE – IT REPLACES BUTTER TO MAKE A MOIST DARK CAKE, TOPPED WITH A BITTER-SWEET STICKY MARMALADE TOPPING.

SERVES EIGHT

INGREDIENTS

- butter, for greasing
- 115g/4oz cooking chocolate (unsweetened), broken into squares
- 3 eggs
- 175g/6oz/scant 1 cup caster (superfine) sugar
- 175ml/6fl oz/¾ cup sour cream
- 200g/7oz/1¾ cups self-raising (self-rising) flour

For the filling and glaze

- 185g/6½oz/⅔ cup bitter orange marmalade
- 115g/4oz plain (semisweet) chocolate, broken into squares
- 60ml/4 tbsp sour cream
- thinly pared and shredded orange rind, to decorate

1 Preheat the oven to 180°C/350°F/ Gas 4. Lightly grease a 1kg/2¼lb loaf tin (pan) with butter, then line the base with a piece of baking parchment. Melt the chocolate in a heatproof bowl over a pan of hot water.

2 Combine the eggs and sugar in a separate mixing bowl. Using a hand-held electric mixer, beat the mixture until it is thick and creamy, then stir in the sour cream and chocolate. Sprinkle over the flour and fold in evenly.

COOK'S TIP

When melting chocolate over hot water, do not allow the base of the bowl to touch the surface of the water, which should be barely simmering.

3 Pour the mixture into the prepared tin and bake for about 1 hour, or until well risen and firm to the touch. Cool for a few minutes in the tin, then turn out on to a wire rack and leave the loaf to cool completely.

4 Make the filling. Spoon two-thirds of the marmalade into a small pan and melt over a low heat. Melt the chocolate in a heatproof bowl over a pan of hot water and stir it into the marmalade with the sour cream.

5 Slice the cake across into three layers and sandwich back together with about half the marmalade filling. Spread the rest evenly over the top of the cake and leave to set. Spoon the remaining marmalade over the cake and sprinkle with shredded orange rind, to decorate.

VARIATION

The filling is also delicious made with other marmalades such as kumquat.

Lemon and Lime Syrup Cake

This cake is perfect for busy cooks as it can be mixed in moments and needs no icing. The simple tangy lime topping transforms it into a fabulously moist cake.

SERVES EIGHT

INGREDIENTS

- 225g/8oz/1 cup butter, softened, plus extra for greasing
- 225g/8oz/2 cups self-raising (self-rising) flour
- 5ml/1 tsp baking powder
- 225g/8oz/generous 1 cup caster (superfine) sugar
- 4 eggs, beaten
- grated rind of 2 lemons
- 30ml/2 tbsp lemon juice

For the topping

- finely pared rind of 1 lime
- juice of 2 limes
- 150g/5oz/¾ cup caster (superfine) sugar

VARIATION

Use lemon rind and juice instead of lime for the topping if you like. You will need only one large lemon.

1 Preheat the oven to 160°C/325°F/Gas 3. Grease and line a 20cm/8in round cake tin (pan). Sift the flour and baking powder into a bowl. Add the caster sugar, butter and eggs and beat until the mixture is smooth and creamy.

2 Beat in the lemon rind and juice. Spoon the mixture into the tin, smooth the surface and gently indent the top with the back of a spoon.

3 Bake for 1¼–1½ hours, or until the cake is golden on top and spongy when lightly pressed, and a skewer inserted in the centre comes out clean.

4 Meanwhile, mix the ingredients for the topping together. As soon as the cake is cooked, remove it from the oven and pour the topping evenly over the surface. Leave the cake to cool fully in the tin before removing and serving.

Moist Orange and Almond Cake

The key to this recipe is to cook the orange slowly first, so that it is completely tender before it is blended. Do not use a microwave to speed things up.

SERVES EIGHT

INGREDIENTS

- 1 large Valencia or Navelina orange
- butter, for greasing
- 3 eggs
- 225g/8oz/generous 1 cup caster (superfine) sugar
- 5ml/1 tsp baking powder
- 225g/8oz/2 cups ground almonds
- 25g/1oz/¼ cup plain (all-purpose) flour
- icing (confectioners') sugar, for dusting

1 Pierce the orange with a skewer. Put it in a deep pan and pour over water to cover it. Bring to the boil, then cover and simmer for 1 hour until the skin is soft. Drain, then cool.

2 Preheat the oven to 180°C/350°F/ Gas 4. Lightly grease a 20cm/8in round cake tin (pan) and line it with baking parchment. Cut the cooled orange in half and discard all the pips (seeds). Place the orange, peel, skin and all, in a food processor or blender and purée until smooth and pulpy.

3 In a bowl, whisk the eggs and sugar until thick. Fold in the baking powder, almonds and flour. Fold in the purée.

4 Pour into the prepared tin, level the surface and bake for 1 hour, or until a skewer inserted into the middle comes out clean. Cool the cake in the tin for 10 minutes, then turn out on to a wire rack, peel off the lining paper and cool completely. Dust the top liberally with icing sugar and serve.

COOK'S TIP

To make a delicious dessert, serve the cake with a little whipped cream and, if you are after extra colour, tuck orange slices underneath it just beforehand. For a special treat, serve the cake with spiced poached kumquats.

LEMON ROULADE WITH LEMON CURD CREAM

This feather-light roulade filled with a rich fresh lemon curd cream makes a marvellous tea-time dessert. Fresh lemon curd is the key to its special flavour, but it can be made ahead, stored in sealed, sterilized jars and kept in the refrigerator.

MAKES EIGHT SLICES

INGREDIENTS

- butter, for greasing
- 4 eggs, separated
- 115g/4oz/generous ½ cup caster (superfine) sugar
- finely grated rind of 2 lemons
- 5ml/1 tsp pure vanilla essence (extract)
- 25g/1oz/¼ cup ground almonds
- 40g/1½oz/⅓ cup plain (all-purpose) flour, sifted
- 45ml/3 tbsp icing (confectioners') sugar, for dusting

For the filling

- 300ml/½ pint/1¼ cups double (heavy) cream
- 60ml/4 tbsp fresh lemon curd

1 Preheat the oven to 190°C/375°F/Gas 5. Grease a 33 × 23cm/13 × 9in Swiss roll tin (jelly roll pan) and line with baking parchment.

2 In a large bowl, beat the egg yolks with half the caster sugar until light and foamy. Beat in the lemon rind and vanilla essence, then lightly fold in the ground almonds and flour using a large metal spoon or spatula.

VARIATION
Substitute orange rind for the roulade and orange curd for the cream filling. Use the rind of three oranges, but the juice of only two to make the fresh orange curd in the same way.

3 Whisk the egg whites until they form stiff peaks. Gradually whisk in the remaining caster sugar to form a stiff meringue. Stir half the mixture into the egg yolk mixture and fold in the rest.

4 Pour into the tin, level the surface with a palette knife (metal spatula) and bake for 10 minutes, or until risen and spongy to the touch. Cover loosely with a sheet of baking parchment and a damp dishtowel. Leave to cool.

5 To make the filling, whip the cream; then lightly fold in the lemon curd.

6 Sift the icing sugar over a piece of baking parchment. Turn the sponge out on to it and trim the edges. Peel off the lining paper and spread the lemon curd cream over the surface of the sponge, leaving a border.

7 Using the paper underneath as a guide, roll up the sponge from one of the long sides. Keep it wrapped in the baking parchment for 1 minute to allow the shape to set. Remove the paper and transfer the roulade to a serving platter, with the seam underneath and serve.

Chocolate Orange Marquise

Here is a cake for people who are passionate about chocolate. The rich, dense flavour is accentuated by fresh orange to make it a truly delectable treat.

SERVES SIX TO EIGHT

INGREDIENTS

- 225g/8oz/1 cup unsalted (sweet) butter, diced, plus extra for greasing
- 200g/7oz/1 cup caster (superfine) sugar
- 60ml/4 tbsp freshly squeezed orange juice
- 350g/12oz dark (bittersweet) chocolate, broken into squares
- 5 eggs
- finely grated rind of 1 orange
- 45ml/3 tbsp plain (all-purpose) flour
- icing (confectioners') sugar and finely pared strips of orange rind, to decorate

COOK'S TIP

It is not necessary to heat the chocolate to melt it. This will make it grainy.

1 Preheat the oven to 180°C/350°F/Gas 4. Grease a 23cm/9in layer cake tin (pan) with a depth of 6cm/2½in. Line the base of the tin with a sheet of baking parchment.

2 Place 90g/3½oz/½ cup of the sugar in a pan. Add the orange juice and stir over a low heat until dissolved.

3 Remove from the heat and stir in the chocolate until melted, then add the butter, piece by piece, until melted.

4 Whisk the eggs with the remaining sugar in a large bowl, until the mixture is pale and very thick. Add the orange rind, then lightly fold the chocolate mixture into the egg mixture. Sift the flour over the top and fold in.

5 Pour the mixture into the prepared tin. Place in a roasting pan, transfer to the oven, then pour hot water into the roasting pan to reach about halfway up the sides of the cake tin.

6 Bake for 1 hour, or until the cake is firm to the touch. Remove the tin from the roasting pan and cool for 20 minutes. Invert the cake on a baking sheet, place a serving plate upside down on top, then carefully turn plate and baking sheet over together so that the cake is transferred to the plate. Dust with a little icing sugar, decorate with strips of orange rind and serve slightly warm or chilled.

SEMOLINA AND NUT HALVA

A traditional semolina pastry from the Eastern Mediterranean, Halva is delicately flavoured with orange, and a fragrant, spicy syrup makes it beautifully moist.

SERVES TEN

INGREDIENTS

- 115g/4oz/½ cup unsalted (sweet) butter, softened
- 115g/4oz/generous ½ cup caster (superfine) sugar
- finely grated rind of 1 orange, plus 30ml/2 tbsp juice
- 3 eggs
- 175g/6oz/1 cup semolina
- 10ml/2 tsp baking powder
- 115g/4oz/1 cup ground hazelnuts
- 50g/2oz/⅓ cup unblanched hazelnuts, toasted and chopped
- 50g/2oz/⅓ cup blanched almonds, toasted and chopped
- thinly pared and shredded rind of 1 orange

For the syrup

- 350g/12oz/1¾ cups caster (superfine) sugar
- 2 cinnamon sticks, halved
- juice of 1 lemon
- 60ml/4 tbsp orange flower water

1 Preheat the oven to 220°C/425°F/Gas 7. Grease and line the base of a deep 23cm/9in square cake tin (pan).

2 Lightly cream the butter in a bowl. Add the sugar, orange rind and juice, eggs, semolina, baking powder and hazelnuts and beat until smooth.

3 Turn into the prepared tin and level the surface. Bake for 20–25 minutes, or until just firm and golden. Remove from the oven and leave to cool in the tin on a wire rack.

4 To make the syrup, put the sugar in a small, heavy pan with 550ml/18fl oz/2½ cups water and the cinnamon sticks. Heat gently, stirring, until the sugar has dissolved completely.

5 Bring to the boil and boil rapidly, without stirring, for 5 minutes. Pour half the syrup into a bowl and add the lemon juice and orange flower water to it. Pour over the halva. Reserve the remainder of the syrup in the pan.

6 Leave the halva in the tin until the syrup is absorbed, then turn it out on to a plate and cut diagonally into diamond-shaped portions. Sprinkle with the nuts.

7 Boil the remaining syrup until slightly thickened, then pour it over the halva. Sprinkle the shredded orange rind over the cake and serve with lightly whipped or clotted cream.

Lemon and Macadamia Bread

The buttery taste of macadamia nuts combines well with the tangy flavour of the lemon rind and yogurt in this delicious bread.

MAKES ONE LOAF

INGREDIENTS

- 40g/1½oz/3 tbsp butter, plus extra for greasing and brushing
- 500g/1¼lb/5 cups unbleached strong white bread flour, plus extra for dusting
- 4ml/¾ tsp salt
- 7.5ml/1½ tsp dried yeast
- 50g/2oz/¼ cup caster (superfine) sugar
- 120ml/4fl oz/½ cup lukewarm milk
- 1 egg
- 175ml/6fl oz/¾ cup lemon yogurt
- 40g/1½oz/¼ cup macadamia nuts, roughly chopped
- 15ml/1 tbsp finely grated lemon rind

1 Line and grease a 15cm/6in cake tin (pan), extending the paper 5cm/2in above the rim, or line and grease a 900g/2lb loaf tin (pan).

2 Sift the flour and salt into a bowl and make a well in the centre.

3 Mix together the yeast, 5ml/1 tsp of the sugar and the lukewarm milk in a bowl and pour into the well. Gradually mix in enough of the surrounding flour to make a thick batter. Sprinkle a little of the remaining flour over the top to cover and leave in a warm place for 20–30 minutes to "sponge" – when the yeast begins to bubble through the flour covering.

4 Add the egg, butter, yogurt and the remaining sugar to the batter, and mix in the remaining flour to make a dough. Knead for 10 minutes, or until smooth and elastic. Cover and leave in a slightly warm place to rise for 1–1½ hours, or until doubled in bulk.

5 Knock back (punch down) the dough and turn out on to a lightly floured surface. Gently knead in the nuts and grated rind. Shape the dough into a ball if using a cake tin, or a long roll if using a loaf tin, and place in the tin. Cover and leave to rise for 30–60 minutes, or until doubled in size.

6 Meanwhile, preheat the oven to 190°C/375°F/Gas 5. Brush with melted butter and bake for 50 minutes, or until golden brown and sounding hollow when tapped. Cool on a wire rack.

COOK'S TIP

Knocking back (punching down) the dough after the first rising helps to disperse the air bubbles produced by the yeast evenly through it. Simply punch it with your fist.

Orange and Coriander Brioches

The warm, spicy flavour of coriander combines particularly well with orange in these dainty, individual brioches, which would make a wonderful mid-morning snack.

MAKES TWELVE

INGREDIENTS

- oil, for greasing
- 225g/8oz/2 cups strong white bread flour
- 10ml/2 tsp easy-blend (rapid-rise) dried yeast
- 2.5ml/½ tsp salt
- 15ml/1 tbsp caster (superfine) sugar
- 10ml/2 tsp coriander seeds, coarsely ground
- grated rind of 1 orange
- 2 eggs, beaten
- 50g/2oz/¼ cup unsalted (sweet) butter, melted
- 1 small egg, beaten, to glaze
- thinly pared and shredded orange rind, to decorate (optional)

1 Grease 12 individual brioche tins (pans). Sift the flour into a bowl and stir in the yeast, salt, sugar, coriander seeds and orange rind. Make a well in the centre, pour in 30ml/2 tbsp lukewarm water, the eggs and melted butter and beat to make a soft dough.

2 Turn on to a lightly floured surface and knead for about 5 minutes, or until smooth and elastic. Return to the clean, oiled bowl, cover with clear film (plastic wrap) and leave in a warm place for 1 hour, or until doubled in bulk.

3 Turn on to a floured surface, knead briefly and roll into a sausage. Cut into 12 pieces. Break off a quarter of each piece. Shape the larger pieces into balls and place in the tins.

4 With the floured handle of a wooden spoon, press a hole in each dough ball. Shape each small piece of dough into a little plug and press into the holes.

COOK'S TIP

These individual brioches look especially attractive if they are made in special brioche tins. However, they can also be made in bun or muffin tins (pans) or as a single loaf in a large brioche tin.

5 Place the brioche tins on a baking sheet. Cover with lightly oiled clear film and leave in a warm place until the dough rises almost to the top of the tins. Preheat the oven to 220°C/425°F/Gas 7. Brush the brioches with beaten egg and bake for 15 minutes, or until golden brown. Sprinkle over extra shreds of orange rind to decorate, if you like, and serve the brioches warm with plenty of butter.

INDEX

Bibliography

The Oxford Companion to Food by Alan Davidson, Oxford University Press, Oxford, 1999

Jane Grigson's Fruit Book by Jane Grigson, Penguin Books, London, 1982

Citrus Varieties of the World by James Saunt, Sinclair International, revised and updated, Norwich, 2000

Picture Acknowledgements

The publishers would like to thank the following companies for the loan of their pictures:

The Art Archive p6 bl

The Mary Evans Picture Library p6 tr

V & A Picture Library p7